A NEW BIOLOGY FOR THE 21ST CENTURY

Committee on a New Biology for the 21st Century:
Ensuring the United States Leads the Coming Biology Revolution

Board on Life Sciences

Division on Earth and Life Studies

NATIONAL RESEARCH COUNCIL
OF THE NATIONAL ACADEMIES

THE NATIONAL ACADEMIES PRESS
Washington, D.C.
www.nap.edu

THE NATIONAL ACADEMIES PRESS   500 Fifth Street, N.W.   Washington, DC 20001

NOTICE: The project that is the subject of this report was approved by the Governing Board of the National Research Council, whose members are drawn from the councils of the National Academy of Sciences, the National Academy of Engineering, and the Institute of Medicine. The members of the committee responsible for the report were chosen for their special competences and with regard for appropriate balance.

This study was supported by the National Institutes of Health through Contract No. N01-OD-4-2139, Task Order 209; the National Science Foundation through Grant No. DBI-0843904; and by the Department of Energy. The content of this publication does not necessarily reflect the views or policies of the sponsoring agencies, nor does the mention of trade names, commercial products, or organizations imply endorsement by the U.S. government.

International Standard Book Number-13:   978-0-309-14488-9 (Book)
International Standard Book Number-10:   0-309-14488-4 (Book)
International Standard Book Number-13:   978-0-309-14486-5 (PDF)
International Standard Book Number-10:   0-309-14486-8 (PDF)
Library of Congress Control Number: 2009939411

Additional copies of this report are available from the National Academies Press, 500 Fifth Street, NW, Lockbox 285, Washington, DC 20055; (800) 624-6242 or (202) 334-3313 (in the Washington metropolitan area); Internet, http://www.nap.edu.

THE NATIONAL ACADEMIES
Advisers to the Nation on Science, Engineering, and Medicine

The **National Academy of Sciences** is a private, nonprofit, self-perpetuating society of distinguished scholars engaged in scientific and engineering research, dedicated to the furtherance of science and technology and to their use for the general welfare. Upon the authority of the charter granted to it by the Congress in 1863, the Academy has a mandate that requires it to advise the federal government on scientific and technical matters. Dr. Ralph J. Cicerone is president of the National Academy of Sciences.

The **National Academy of Engineering** was established in 1964, under the charter of the National Academy of Sciences, as a parallel organization of outstanding engineers. It is autonomous in its administration and in the selection of its members, sharing with the National Academy of Sciences the responsibility for advising the federal government. The National Academy of Engineering also sponsors engineering programs aimed at meeting national needs, encourages education and research, and recognizes the superior achievements of engineers. Dr. Charles M. Vest is president of the National Academy of Engineering.

The **Institute of Medicine** was established in 1970 by the National Academy of Sciences to secure the services of eminent members of appropriate professions in the examination of policy matters pertaining to the health of the public. The Institute acts under the responsibility given to the National Academy of Sciences by its congressional charter to be an adviser to the federal government and, upon its own initiative, to identify issues of medical care, research, and education. Dr. Harvey V. Fineberg is president of the Institute of Medicine.

The **National Research Council** was organized by the National Academy of Sciences in 1916 to associate the broad community of science and technology with the Academy's purposes of furthering knowledge and advising the federal government. Functioning in accordance with general policies determined by the Academy, the Council has become the principal operating agency of both the National Academy of Sciences and the National Academy of Engineering in providing services to the government, the public, and the scientific and engineering communities. The Council is administered jointly by both Academies and the Institute of Medicine. Dr. Ralph J. Cicerone and Dr. Charles M. Vest are chair and vice chair, respectively, of the National Research Council.

www.national-academies.org

COMMITTEE ON A NEW BIOLOGY FOR THE 21ST CENTURY: ENSURING THE UNITED STATES LEADS THE COMING BIOLOGY REVOLUTION

THOMAS CONNELLY (Cochair), DuPont Company, Wilmington, Delaware
PHILLIP SHARP (Cochair), Massachusetts Institute of Technology, Cambridge
DENNIS AUSIELLO, Massachusetts General Hospital and Harvard Medical School, Boston
MARIANNE BRONNER-FRASER, California Institute of Technology, Pasadena
INGRID BURKE, University of Wyoming, Laramie
JOHN BURRIS, Burroughs Wellcome Fund, Research Triangle Park, North Carolina
JONATHAN EISEN, University of California, Davis
ANTHONY JANETOS, Joint Global Change Research Institute, College Park, Maryland
RICHARD KARP, International Computer Science Institute and University of California, Berkeley
PETER KIM, Merck Research Laboratories, North Wales, Pennsylvania
DOUGLAS LAUFFENBURGER, Massachusetts Institute of Technology, Cambridge
MARY LIDSTROM, University of Washington, Seattle
WENDELL LIM, University of California, San Francisco
MARGARET MCFALL-NGAI, University of Wisconsin, Madison
ELLIOT MEYEROWITZ, California Institute of Technology, Pasadena
KEITH YAMAMOTO, University of California, San Francisco

Staff

ANN REID, Study Director, Board on Life Sciences
AMANDA CLINE, Senior Program Assistant, Board on Life Sciences
FRANCES SHARPLES, Director, Board on Life Sciences
SANJAY MAGAVI, Christine Mirzayan Fellow, Board on Life Sciences
KATHERINE SAYLOR, Christine Mirzayan Fellow, Board on Life Sciences

Preface

Biological research is in the midst of a revolutionary change due to the integration of powerful technologies along with new concepts and methods derived from inclusion of physical sciences, mathematics, computational sciences, and engineering. As never before, advances in biological sciences hold tremendous promise for surmounting many of the major challenges confronting the United States and the world. Historically, major advances in science have provided solutions to economic and social challenges. At the same time, those challenges have inspired science to focus its attention on critical needs. Scientific efforts based on meeting societal needs have laid the foundation for countless new products, industries, even entire economic sectors that were unimagined when the work began.

The lessons of history led the Committee on a New Biology for the 21st Century to recommend that a New Biology Initiative be put in place and charged with finding solutions to major societal needs: sustainable food production, protection of the environment, renewable energy, and improvement in human health. These challenges represent both the mechanism for accelerating the emergence of a New Biology and its first fruits. Responding to its Statement of Task, the committee found the answer to the question "How can a fundamental understanding of living systems reduce uncertainty about the future of life on earth, improve human health and welfare, and lead to the wise stewardship of our planet?" in calling for a national initiative to apply the potential of the New Biology to addressing these societal challenges.

As the report explains, the essence of the New Biology is integration— re-integration of the many subdisciplines of biology, and the integration into biology of physicists, chemists, computer scientists, engineers, and mathematicians to create a research community with the capacity to tackle a broad range of scientific and societal problems. The committee chose biological approaches to solving problems in the areas of food, environment, energy and health as the most inspiring goals to drive the development of the New Biology. But these are not the only problems that we both hope and expect a thriving New Biology to be able to address; fundamental questions in all areas of biology, from understanding the brain to carbon cycling in the ocean, will all be more tractable as the New Biology grows into a flourishing reality. Given the fundamental unity of biology, it is our hope and our expectation that the New Biology will contribute to advances across the life sciences. Throughout the report, "New Biology" is capitalized to emphasize that it is intended to be an additional and complementary effort to traditional life sciences research, not a replacement. Peer-reviewed, independent investigator-initiated research is the foundation on which the New Biology rests and on which it will continue to rely.

Many exciting and important areas of biological research are not considered in this report. America's research capability in life sciences leads the world. This committee strongly endorses current research endeavors, both in the public and private sector. Within biology, the excellent work underway must be continued. But for this study, the intent was not to comprehensively review all life sciences research. Instead the committee focused on those opportunities that cannot be addressed by any one subdiscipline or agency—opportunities that require integration across biology and with other sciences and engineering, and that are difficult to capitalize on within traditional institutional and funding structures.

It is not merely the sciences that need to be integrated. The New Biology will draw on the research and development capabilities of universities, government, and industry. Individual federal agencies will continue to lead important, independent efforts. For the New Biology to flourish, however, interagency co-leadership of projects will be needed to a far greater extent than is the case today. This approach is not simply a matter of funding. The combined capabilities and expertise of numerous organizations are required to address society's greatest challenges.

This study represents the collective efforts of the committee during meetings, workshops, a December 2008 Biology Summit, and many teleconferences. We would like to thank the Summit and workshop participants for their valuable input. We also thank the committee members who volunteered countless hours and the Board of Life Sciences staff for their efforts and dedication to the study.

America's investment in basic research in the life sciences has paid rich dividends. A commitment to the New Biology will extend this proud record.

In the words of President Obama when he addressed the 2009 annual meeting of the National Academy of Sciences:

As you know, scientific discovery takes far more that the occasional flash of brilliance—as important as that can be. Usually, it takes time and hard work and patience; it takes training; it requires the support of a nation. But it holds promise like no other area of human endeavor.

The well-being, security, and prosperity of our nation are the prize. We fully endorse the recommendations here presented.

THOMAS CONNELLY
PHILLIP SHARP
Co-chairs
Committee on a New Biology for the 21st Century:
Ensuring the United States Leads the Coming
Biology Revolution

Acknowledgments

This report has been reviewed in draft form by persons chosen for their diverse perspectives and technical expertise in accordance with procedures approved by the National Research Council's Report Review Committee. The purpose of the independent review is to provide candid and critical comments that will assist the institution in making the published report as sound as possible and to ensure that the report meets institutional standards of objectivity, evidence, and responsiveness to the study charge. The review comments and draft manuscript remain confidential to protect the integrity of the deliberative process. We wish to thank the following for their review of the report:

Frances H. Arnold, California Institute of Technology, Pasadena
Ann M. Arvin, Stanford University, California
David Baltimore, California Institute of Technology, Pasadena
Floyd E. Bloom, The Scripps Research Institute, La Jolla, California
Jeff Dangl, University of North Carolina, Chapel Hill
Susan Desmond-Hellmann, University of California, San Francisco
Mark Ellisman, University of California, San Diego
Paul Falkowski, Rutgers University, New Brunswick, New Jersey
Adam Godzik, Burnham Institute for Medical Research, La Jolla, California
David Goldston, Princeton University, New Jersey
James Hanken, Harvard University, Cambridge, Massachusetts
Robert Langer, Massachusetts Institute of Technology, Cambridge
Rick Miranda, Colorado State University, Fort Collins
Norman Pace, University of Colorado, Boulder
Camille Parmesan, University of Texas, Austin
Peter H. Raven, Missouri Botanical Garden, St. Louis

Gene Robinson, University of Illinois, Urbana
Bruce W. Stillman, Cold Spring Harbor Laboratory, New York

Although the reviewers listed above have provided constructive comments and suggestions, they were not asked to endorse the conclusions or recommendations, nor did they see the final draft of the report before its release. The review of the report was overseen by **Marvalee H. Wake** (University of California, Berkeley) and **John Dowling** (Harvard University, Cambridge, Massachusetts). Appointed by the National Research Council, they were responsible for making certain that an independent examination of this report was carried out in accordance with institutional procedures and that all review comments were carefully considered. Responsibility for the final content of the report rests entirely with the authoring committee and the institution.

The committee benefited from discussions with several speakers, whom we would like to thank for their help. At its first meeting, on November 4, 2008, the committee met with: **Ralph Cicerone,** President, National Academy of Sciences, **Charles M. Vest**, President, National Academy of Engineering, **James Jensen**, Director of Congressional and Government Affairs, National Academies, **William Bonvillian**, Director, Massachusetts Institute of Technology, Washington, **Patrick White**, Director of Federal Relations, Association of American Universities, **Howard Minigh**, President and CEO, CropLife International, **John Pierce**, Vice President, DuPont Applied BioSciences–Technology, and **Anthony Janetos**, Director, Joint Global Change Research Institute, College Park, Maryland.

We also thank **Robert Lue**, Harvard University, **Timothy J. Donohue**, University of Wisconsin-Madison, **William K. Lauenroth**, University of Wyoming, for helpful discussions, **Joshua V. Troll**, University of Wisconsin, Madison and **Charina Choi**, University of California, San Francisco for contributing figures, and **Steve Olson** and **Paula Tarnapol Whitacre** for writing and editing assistance.

Contents

Summary

In July, 2008, the National Institutes of Health (NIH), National Science Foundation (NSF), and Department of Energy (DOE) asked the National Research Council's Board on Life Sciences to convene a committee *to examine the current state of biological research in the United States and recommend how best to capitalize on recent technological and scientific advances that have allowed biologists to integrate biological research findings, collect and interpret vastly increased amounts of data, and predict the behavior of complex biological systems.* From September 2008 through July of 2009, a committee of 16 experts from the fields of biology, engineering and computational science undertook to delineate those scientific and technological advances and come to a consensus on how the U.S. might best capitalize on them. This report, authored by the Committee on a New Biology for the 21st Century, describes the committee's work and conclusions.

The committee concluded that biological research has indeed experienced extraordinary scientific and technological advances in recent years. In the chapter entitled "Why Now?" the committee describes the integration taking place within the field of biology, the increasingly fruitful collaboration of biologists with scientists and engineers from other disciplines, the technological advances that have allowed biologists to collect and make sense of ever more detailed observations at ever smaller time intervals, and the enormous and largely unanticipated payoffs of the Human Genome Project. Despite this potential, the challenge of advancing from identifying parts, to defining complex systems, to systems design, manipulation, and prediction is still well beyond current capabilities, and the barriers to advancement are similar at all levels from cells to ecosystems.

Having delineated the advances, the committee set about reaching an agreement as to how the U.S. could best capitalize on them. The committee was invited to use the following series of questions to guide its discussions:

1

- *What fundamental biological questions are ready for major advances in understanding? What would be the practical result of answering those questions? How could answers to those questions lead to high impact applications in the near future?*
- *How can a fundamental understanding of living systems reduce uncertainty about the future of life on earth, improve human health and welfare, and lead to the wise stewardship of our planet? Can the consequences of environmental, stochastic or genetic changes be understood in terms of the related properties of robustness and fragility inherent in all biological systems?*
- *How can federal agencies more effectively leverage their investments in biological research and education to address complex problems across scales of analysis from basic to applied? In what areas would near term investment be most likely to lead to substantial long-term benefit and a strong, competitive advantage for the United States? Are there high-risk, high pay-off areas that deserve serious consideration for seed funding?*
- *Are new funding mechanisms needed to encourage and support cross-cutting, interdisciplinary or applied biology research?*
- *What are the major impediments to achieving a newly integrated biology?*
- *What are the implications of a newly integrated biology for infrastructural needs?*
- *How should infrastructural priorities be identified and planned for?*
- *What are the implications for the life sciences research culture of a newly integrated approach to biology? How can physicists, chemists, mathematicians and engineers be encouraged to help build a wider biological enterprise with the scope and expertise to address a broad range of scientific and societal problems?*
- *Are changes needed in biology education—to ensure that biology majors are equipped to work across traditional subdisciplinary boundaries, to provide biology curricula that equip physical scientists and engineers to take advantage of advances in biological science, and to provide nonscientists with a level of biological understanding that gives them an informed voice regarding relevant policy proposals? Are alternative degree programs needed or can biology departments be organized to attract and train students able to work comfortably across disciplinary boundaries?*

The committee found that the third bullet, "*How can federal agencies more effectively leverage their investments in biological research and education to address complex problems across scales of analysis from basic to applied? In what areas would near term investment be most likely to lead to substantial long-term benefit and a strong, competitive advantage for the United States?*" provided a compelling platform from which to consider each of the questions, and a robust framework upon which to organize its conclusions. Thus, the committee's overarching recommendation is that the most effective leveraging of investments would come from a coordinated, interagency effort to encourage the emergence of a New Biology approach that would enunciate and address broad and challenging societal problems. The committee focused on examples of opportunities that cannot be addressed by any one subdiscipline or agency—opportunities that require integration across biology and with other sciences and engineering, and that are difficult to capitalize on within traditional institutional and funding structures. Fully realizing these opportunities will require the enabling of an integrated approach to biological research, an approach the committee calls the New Biology.

The essence of the New Biology, as defined by the committee, is integration—re-integration of the many sub-disciplines of biology, and the integration into biology of physicists, chemists, computer scientists, engineers, and mathematicians to create a research community with the capacity to tackle a broad range of scientific and societal problems. Integrating knowledge from many disciplines will permit deeper understanding of biological systems, which will both lead to biology-based solutions to societal problems and also feed back to enrich the individual scientific disciplines that contribute new insights. The New Biology is not intended to replace the research that is going on now; that research, much of it fundamental and curiosity-driven by individual scientists, is the foundation on which the New Biology rests and on which it will continue to rely.

Instead, the New Biology represents an additional, complementary approach to biological research. Purposefully organized around problem-solving, this approach marshals the basic research to advance fundamental understanding, brings together researchers with different expertise, develops the technologies required for the task and coordinates efforts to ensure that gaps are filled, problems solved, and resources brought to bear at the right time. Combining the strengths of different communities does not necessarily mean bringing these experts into the same facility to work on one large project—indeed, advanced communication and informatics infrastructures make it easier than ever to assemble virtual collaborations at different scales. The New Biology approach would aim to attract the best minds from across the scientific landscape to particular problems, ensure that innovations and advances are swiftly communicated, and provide the tools and technologies needed to succeed. The committee expects that such efforts would include projects at different scales, from individual laboratories, to collaborations involving many participants, to consortia involving multiple institutions and types of research.

Many scientists in the United States are already practicing the integrated and interdisciplinary approach to biology that the committee has called the New Biology. The New Biology is indeed already emerging, but it is as yet poorly recognized, inadequately supported, and delivering only a fraction of its potential. The committee concludes that the most effective way to speed the emergence of the New Biology is to challenge the scientific community to discover solutions to major societal problems. In the chapter entitled "How the New Biology Can Address Societal Challenges" the committee describes four broad challenges, in food, environment, energy and health that could be tackled by the New Biology. These challenges represent both the mechanism for accelerating the emergence of a New Biology and its first fruits. The committee chose to focus on these four areas of societal need because the benefits of achieving these goals would be large, progress would be assessable, and both the scientific community and the public would find such goals inspirational. Each challenge will require technological and conceptual advances that are not now at hand, across a disciplinary spectrum that is not now encompassed by the field. Achieving these goals will demand, in each case, transformative advances. It can be argued, however, that other challenges could serve the same purpose. Large-scale efforts to understand how the first cell came to be, how the human brain works, or how living organisms affect the cycling of carbon in the ocean could also drive the development of the New Biology and of the technologies and sciences necessary to advance the entire field. In the committee's view, one of the most exciting aspects of the New Biology Initiative is that success in achieving the four goals chosen here as examples will propel advances in fundamental understanding throughout the life sciences. Because biological systems have so many fundamental similarities, the same technologies and sciences developed to address these four challenges will expand the capabilities of all biologists.

1. Generate food plants to adapt and grow sustainably in changing environments
　　The New Biology could deliver a dramatically more efficient approach to developing plant varieties that can be grown sustainably under local conditions. The result of this focused and integrated effort will be a body of knowledge, new tools, technologies, and approaches that will make it possible to adapt all sorts of crop plants for efficient production under different conditions, a critical contribution toward making it possible to feed people around the world with abundant, healthful food, adapted to grow efficiently in many different and ever-changing local environments.

2. Understand and sustain ecosystem function and biodiversity in the face of rapid change
　　Fundamental advances in knowledge and a new generation of tools and technologies are needed to understand how ecosystems function, measure ecosystem services, allow restoration of damaged ecosystems, and minimize harmful impacts of human activities and climate change. What is needed is the New Biology, combining the knowledge base of ecology with those of organismal biology, evolutionary and comparative biology, climatology, hydrology, soil science, and environmental, civil, and systems engineering, through the unifying languages of mathematics, modeling, and computational science. This integration has the potential to generate breakthroughs in our ability to monitor ecosystem function, identify ecosystems at risk, and develop effective interventions to protect and restore ecosystem function.

3. Expand sustainable alternatives to fossil fuels

Making efficient use of plant materials—biomass—to make biofuels is a systems challenge, and this is another example of an area where the New Biology can make a critical contribution. At its simplest, the system consists of a plant that serves as the source of cellulose and an industrial process that turns the cellulose into a useful product. There are many points in the system that can be optimized. The New Biology offers the possibility of advancing the fundamental knowledge, tools, and technology needed to optimize the system by tackling the challenge in a comprehensive way.

4. Understand individual health

The goal of a New Biology approach to health is to make it possible to monitor each individual's health and treat any malfunction in a manner that is tailored to that individual. In other words, the goal is to provide individually predictive surveillance and care. Between the starting point of an individual's genome sequence and the endpoint of that individual's health is a web of interacting networks of staggering complexity. The New Biology can accelerate fundamental understanding of the systems that underlie health and the development of the tools and technologies that will in turn lead to more efficient approaches to developing therapeutics and enabling individualized, predictive medicine.

Finally, in the chapter entitled "Putting the New Biology to Work," the committee proposes that a national initiative dedicated to addressing challenges like those described for the areas of food, the environment, energy, and health would provide a framework whereby the U.S. could best capitalize on recent scientific and technological advances. The committee recommends setting big goals and then letting the problems drive the science. It contends that interagency collaboration will be essential and that information technologies will be of central importance. Finally, the committee discusses new approaches to education that could speed the emergence of the New Biology, and provides examples of how a national initiative could spur the implementation of those new approaches.

The committee does not provide a detailed plan for implementation of such a national initiative, which would depend strongly on where administrative responsibility for the initiative is placed. Should the concept of an initiative be adopted, the next step would be careful development of strategic visions for the programs and a tactical plan with goals. It would be necessary to identify imaginative leaders, carefully map the route from 'grand visions' to specific programs, and develop ambitious, but measurable milestones, ensuring that each step involves activities that result in new knowledge and facilitates the smooth integration of cooperative interdisciplinary research into the traditional research culture.

A New Biology Initiative would represent a daring addition to the nation's research portfolio, but the potential benefits are remarkable and far-reaching: a life sciences research community engaged in the full spectrum of knowledge discovery and its application; new bio-based industries; and most importantly, innovative means to produce food and biofuels sustainably, monitor and restore ecosystems and improve human health. To that end, the committee provides the following findings and recommendations:

Finding 1

• The United States and the world face serious societal challenges in the areas of food, environment, energy, and health.

• Innovations in biology can lead to sustainable solutions for all of these challenges. Solutions in all four areas will be driven by advances in fundamental understanding of basic biological processes.

• For each of these challenges, solutions are within reach, based on building the capacity to understand, predict, and influence the responses and capabilities of complex biological systems.

• There is broad support across the scientific community for pursuing interdisciplinary research, but opportunities to do so are constrained by institutional barriers and available resources.

• Approaches that integrate a wide range of scientific disciplines, and draw on the strengths and resources of universities, federal agencies, and the private sector will accelerate progress toward making this potential a reality.

• The best way for the United States to capitalize on this scientific and technological opportunity is to add to its current research portfolio a New Biology effort that will accelerate understanding of complex biological systems, driving rapid progress in meeting societal challenges and advancing fundamental knowledge.

Recommendation 1
The committee recommends a national initiative to accelerate the emergence and growth of the New Biology to achieve solutions to societal challenges in food, energy, environment, and health.

Finding 2

• For its success, the New Biology will require the creative drive and deep knowledge base of individual scientists from across biology and many other disciplines including physical, computational and geosciences, mathematics, and engineering.

• The New Biology offers the potential to address questions at a scale and with a focus that cannot be undertaken by any single scientific community, agency or sector.

• Providing a framework for different communities to work together will lead to synergies and new approaches that no single community could have achieved alone.

• A broad array of programs to identify, support, and facilitate biology research exists in the federal government but value is being lost by not integrating these efforts.

• Interagency insight and oversight is critical to support the emergence and growth of the New Biology Initiative. Interagency leadership will be needed to oversee and coordinate the implementation of the initiative, evaluate its progress, establish necessary working subgroups, maintain communication, guard against redundancy, and identify gaps and opportunities for leveraging results across projects.

Recommendation 2:
The committee recommends that the national New Biology Initiative be an interagency effort, that it have a timeline of at least 10 years, and that its funding be in addition to current research budgets.

Finding 3
• Information is the fundamental currency of the New Biology.
• Solutions to the challenges of standardization, exchange, storage, security, analysis, and visualization of biological information will multiply the value of the research currently being supported across the federal government.
• Biological data are extraordinarily heterogeneous and interrelating various bodies of data is currently precluded by the lack of the necessary information infrastructure.
• It is critical that all researchers be able to share and access each others' information in a common or fully interactive format.
• The productivity of biological research will increasingly depend on long-term, predictable support for a high-performance information infrastructure.

Recommendation 3
The committee recommends that, within the national New Biology Initiative, priority be given to the development of the information technologies and sciences that will be critical to the success of the New Biology.

Finding 4
• Investment in education is essential if the new biology is to reach its full potential in meeting the core challenges of the 21st century.
• The New Biology Initiative provides an opportunity to attract students to science who want to solve real-world problems.
• The New Biologist is not a scientist who knows a little bit about all disciplines, but a scientist with deep knowledge in one discipline and a "working fluency" in several.

• Highly developed quantitative skills will be increasingly important.

• Development and implementation of genuinely interdisciplinary undergraduate courses and curricula will both prepare students for careers as New Biology researchers and educate a new generation of science teachers well versed in New Biology approaches.

• Graduate training programs that include opportunities for interdisciplinary work are essential.

• Programs to support faculty in developing new curricula will have a multiplying effect.

Recommendation 4
The committee recommends that the New Biology Initiative devote resources to programs that support the creation and implementation of interdisciplinary curricula, graduate training programs, and educator training needed to create and support New Biologists.

Introduction

A Vision of the Future

Imagine a world:

- **where there is abundant, healthful food for everyone**
- **where the environment is resilient and flourishing**
- **where there is sustainable, clean energy**
- **where good health is the norm**

Each of these goals is a daunting challenge. Furthermore, none can be attained independently of the others—we want to grow more food without using more energy or harming natural environments, and we want new sources of energy that do not contribute to global warming or have adverse health effects. The problems raised by these fundamental biological and environmental questions are interdependent and "solutions" that work at cross purposes will not in fact be solutions.

Fortunately, advances in the life sciences have the potential to contribute innovative and mutually reinforcing solutions to reach all of these goals and, at the same time, serve as the basis for new industries that will anchor the economies of the future. Here are some of the many different ways in which the life sciences could contribute to meeting these challenges:

- A wide variety of plants with faster maturation, drought tolerance, and disease resistance could contribute to a sustainable increase in local food production.
- Food crops could be engineered for higher nutritional value, including higher concentrations of vitamins and healthier oils.
- Critical habitats could be monitored by arrays of remote sensors,

enabling early detection of habitat damage and providing feedback on the progress of restoration efforts.

• Water supplies and other natural resources could be monitored and managed using biosensors and other biologically based processes.

• Biological systems could remove more carbon dioxide from the atmosphere, thus helping to maintain a stable climate; the carbon they capture could be used to create biologically based materials for construction and manufacturing.

• Biological sources could contribute at least 20 percent of the fuel for transportation through a 10-fold increase in biofuel production.

• Bio-inspired approaches to producing hydrogen could provide another affordable and sustainable source of fuel.

• Biologically inspired approaches to capturing solar energy could increase the efficiency and lower the cost of photovoltaic technology.

• Manufactured products could increasingly be made from renewable resources and be either recyclable or biodegradable.

• Industrial manufacturing processes could be designed to produce zero waste through a combination of biological treatment of byproducts and efficient recycling of water and other manufacturing inputs.

• Greater understanding of what it means to be healthy could lead to health care focused on maintaining health rather than reacting to illness.

• Individualized risk profiles and early detection could make it possible to provide each person with the right care at the right time.

Science and technology alone, of course, cannot solve all of our food, energy, environmental, and health problems. Political, social, economic, and many other factors have major roles to play in both setting and meeting goals in these areas. Indeed, increased collaboration between life scientists and social scientists is another exciting interface that has much to contribute to developing and implementing practical solutions. But the life sciences have the potential to provide a set of tools and solutions that can significantly increase the options available to society for dealing with problems. Integration of the biological sciences with physical and computational sciences, mathematics, and engineering promises to build a wider biological enterprise with the scope and expertise to address a broad range of scientific and societal problems. The following chapters will discuss why the life sciences are poised to tackle major challenges of the 21st century, describe why we reside at such an exceptional moment for the life sciences, and finally, provide recommendations for shaping investment in life science research.

1

The New Biology's Great Potential

In July, 2008, the National Institutes of Health (NIH), National Science Foundation (NSF), and Department of Energy (DOE) asked the National Research Council's Board on Life Sciences to examine the current state of biological research in the United States and recommend how best to capitalize on recent technological and scientific advances that have allowed biologists to integrate biological research findings, collect and interpret vastly increased amounts of data, and predict the behavior of complex biological systems. The board convened a committee entitled the Committee on a New Biology for the 21st Century to take on this assignment. The committee's statement of task was broad, calling for an appraisal of areas in which the life sciences are poised to make major advances and how these advances could contribute to practical applications and improved environmental stewardship, human health, and quality of life. The committee was asked to examine current trends toward integration and synthesis within the life sciences and the increasingly important role of interdisciplinary teams and the resultant implications for funding strategies, decision-making, infrastructure, and education in the life sciences.

Ultimately, the committee was asked to make recommendations aimed at ensuring that the United States takes the lead in the emergence of a biological science that will support a higher level of confidence in our understanding of living systems, thus reducing uncertainty about the future, contributing to innovative solutions for practical problems, and allowing the development of robust and sustainable new technologies.

The study included a "Biology Summit" on December 3, 2008, at which leaders of major biology research funding agencies and private research foundations outlined the great potential of biology research and the challenges in reaching that potential. Other speakers included private sector "consumers" of life sciences research results, a university president, and several biology researchers who illustrated the interdisciplinary, high-impact biology research that is already taking place. The Summit proceedings were published as a workshop report in January 2009 (National Research Council, 2009).

Given the statement of task's imperative that the committee provide recommendations to federal agencies on how best to support emerging capabilities in the life sciences, the committee invited several speakers to its first meeting in November 2008 for advice on how to develop effective and implementable recommendations. Both Ralph Cicerone, president of the National Academy of Sciences, and Charles Vest, president of the National Academy of Engineering, spoke with the committee, as did several speakers knowledgeable about the impact of past National Academies reports. The committee also heard talks on how basic life sciences research contributes to diverse economic sectors and international science policy efforts. After the Biology Summit in December 2008, the committee met three times, in February, April, and July 2009, to develop the report and its recommendations.

Because the statement of task was broad, the committee wrestled with how best to address it. Some of the questions the committee was invited to consider focused on scientific priorities—for example, what fundamental biological questions are ready for major advances in understanding? Other questions were more practical—are new funding mechanisms needed to support cross-cutting, interdisciplinary, or applied biology research? The committee explored several different approaches to addressing so wide a range of questions. One approach would have been to examine the current life sciences landscape and highlight specific areas of biological research that are particularly exciting or promising. This was the approach taken in a 1989 National Research Council (NRC) report called *Opportunities in Biology* (National Research Council, 1989). Over 400 pages in length and with a chapter devoted to each of nine major subdisciplines of biology, the report identified questions each of those sub-disciplines was poised to answer. Certainly there would be no shortage of material if this committee had followed that model: Across biology from neuroscience to organismal biology to ecology, genomics, and bioengineering, the pace of discovery is rapid, making ambitious goals ever more realistic (Institute of Medicine, 2008; Schwenk et al., 2009; National Academy of Engineering, 2009). But such a list would, by necessity, be incomplete and almost immediately outdated. Furthermore, the committee felt that such an approach would miss a critical insight with tremendous implications.

Biology is at a point of inflection. Years of research have generated detailed information about the components of the complex systems that characterize life—genes, cells, organisms, ecosystems—and this knowledge has begun to fuse into greater understanding of how all those components work together as systems. Powerful tools are allowing biologists to probe complex systems in ever-greater detail, from molecular events in individual cells to global biogeochemical cycles. Integration within biology and increasingly fruitful collaboration with physical, earth, and computational scientists, mathematicians, and engineers are making it possible to predict and control the activities of biological systems in ever greater detail.

These trends both reflect and depend on the fundamental nature of life. Biology's tremendous potential rests on two powerful facts, the first being that all organisms are related by evolution. Therefore, work on one gene, one cell, one species is directly relevant to understanding all others because processes may be identical or highly similar between different organisms due to their shared descent. Second, the process of evolution has generated countless variations on these common themes—a vast array of organisms with myriad adaptations to diverse environments—and comparison is a powerful illuminator. Biology is now at a point of being able to capitalize on these essential characteristics of the living world, and that ability has implications across many sectors. Just as the Internet, combined with powerful search engines, makes vast amounts of information accessible, the core commonalities of biology, combined with increasingly sophisticated ways to compare, predict, and manipulate their characteristics, can make the resources of biology accessible for a wide range of applications. The committee concluded that the life sciences have reached a point where a new level of inquiry is possible, a level that builds on the strengths of the traditional research establishment but provides a framework to draw on those strengths and focus them on large questions whose answers would provide many practical benefits. We call this new level of inquiry the New Biology and believe that it has the potential to take on more ambitious challenges than ever before. As examples of the kinds of challenges this approach can address, the committee has chosen aspects of critical economic sectors—food, the environment, energy, and health—to which the New Biology could make important contributions. Though the problems are indeed diverse, many of the solutions the life sciences can offer will derive from greater understanding of core biological processes— processes that are common to all living systems. Achieving understanding at this systemic level is the promise of the New Biology.

Biological research is supported by many federal agencies (Box 1.1). Each nurtures a talented community of scientists and engineers, supports technology and tool development, builds infrastructure, and funds training and education programs. Because of biology's increasing trend toward integration, the work of these agencies is potentially more complementary than ever before. In fact, the committee concludes that if a framework were in place for these agencies and others to work together and solicit input from academia, the private sector, and foundations, significant progress could be made on meeting major societal challenges.

The committee concludes that a bold proposal to focus the newly emerging capabilities of biological research on major societal challenges is timely and that a relatively small investment could have large benefits by leveraging resources and skills across the federal government, private, and academic sectors.

BOX 1.1
Federal Departments and Agencies that
Support Biology Research

Department of Agriculture (USDA)
    Agricultural Research Service (ARS)
    Cooperative State Research, Education and Extension Service (CSREES)
    Forest Service (FS)

Department of Commerce (DOC)
    National Oceanic and Atmospheric Administration (NOAA)
    National Marine Fisheries Service (NMFS)
    National Institute of Standards and Technology (NIST)

Department of Defense (DOD)
    Defense Advanced Research Projects Agency (DARPA)
    Defense Science and Technology Program
    Office of Naval Research (ONR)
    U.S. Army Medical Research and Materiel Command (USAMRMC)

Department of Energy (DOE)
    Science Office
    National Laboratories

Department of Homeland Security (DHS)

Department of the Interior (DOI)
    Fish and Wildlife Service (FWS)
    Geological Survey (USGS)

Environmental Protection Agency (EPA)

Health and Human Services Department (HHS)
    Centers for Disease Control and Prevention (CDC)
    Food and Drug Administration (FDA)
    National Institutes of Health (NIH)

National Aeronautics and Space Administration (NASA)

National Science Foundation (NSF)

Veterans Affairs Department (VA)

The committee discussed each of the points in its statement of task (Appendix A). Although each question may not be explicitly addressed in this report, those discussions had a major impact on leading the committee to recommend a problem-focused approach. For example, the committee argues that focusing attention on solving practical problems will require, and in turn lead to, advances in fundamental understanding. Implications for infrastructure, education, and research culture are raised throughout the report, and suggestions are offered for positive approaches to implement change. It is the committee's hope that the report presents a convincing vision of how federal agencies could more effectively leverage their investments in biological research and education to address complex problems and a compelling argument that this near-term investment will lead to substantial long-term benefits and a strong, competitive advantage for the United States.

2

How the New Biology
Can Address Societal Challenges

In the 1800s, those who studied the living world were called "naturalists" and they were highly interdisciplinary, combining observations from biology, geology, and physics to describe the natural world. In this 200th anniversary year of Darwin's birth, after decades of highly productive specialization, the study of life is again becoming more interdisciplinary, by necessity combining previously disparate fields to create a "New Biology." The essence of the New Biology is re-integration of the subdisciplines of biology, along with greater integration with the physical and computational sciences, mathematics, and engineering in order to devise new approaches that tackle traditional and systems level questions in new, interdisciplinary, and especially, quantitative ways (Figure 2.1).

As illustrated in Figure 2.1, the New Biology relies on integrating knowledge from many disciplines to derive deeper understanding of biological systems. That deeper understanding both allows the development of biology-based solutions for societal problems and also feeds back to enrich the individual scientific disciplines that contributed to the new insights. It is critically important to recognize that the New Biology does not replace the research that is going on now; that research is the foundation on which the New Biology rests and on which it will continue to rely. If we compare our understanding of the living world to the assembly of a massive jigsaw puzzle, each of the subdisciplines of biology has been assembling sections of the puzzle. The individual sections are far from complete and continued work to fill those gaps is critical. Indeed, biological systems are so complex that it is likely that major new discoveries are still to be expected, and new discoveries very frequently come from individual scientists who make the intellectual leap from the particular system they study to an insight that illuminates many biological processes. The additional contri-

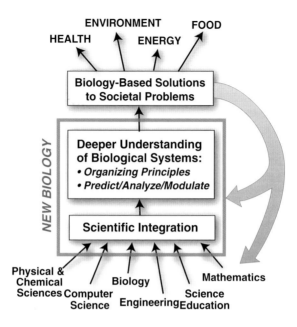

FIGURE 2.1 What is the New Biology?
SOURCE: Committee on a New Biology for the 21st Century.

bution of the New Biology is to focus on the connections between the partially assembled puzzle sections and dramatically speed up overall assembly.

WHO IS THE NEW BIOLOGIST?

The committee believes that virtually every biologist who reads this report's description of the New Biologist will recognize him or herself. All biologists think across levels of biological complexity—molecular biologists consider the impact of genetic regulatory pathways on the health of organisms, ecologists consider the impact of environmental change on the gene pool of an ecosystem, and neuroscientists link cell-to-cell communication with behavior. Rare is the biologist who does not use computational tools to analyze data, or rely on large-scale shared facilities for some experiments. And an increasing fraction of biologists collaborate closely with physical scientists, computational scientists or engineers. The workshop held at the beginning of this committee's work highlighted a number of laboratories where the New Biology is already well advanced (Box 2.1). The committee does not intend to suggest that there is a stark division between 'old' biologists and 'new' biologists, but rather that there is a continuum from more reductionist, focused research within particular

BOX 2.1
A Wiring Diagram for Cells

Cellular systems can be represented in "wiring diagrams" analogous to those of electronic circuits. But the components in the diagram are proteins, nucleic acids, and other biologically active molecules while the wires are interactions among those components.

Lucy Shapiro's laboratory at the Stanford University School of Medicine chose a simple organism, a bacterium called *Caulobacter crescentus*, and set out to understand all the integrated processes that this organism needs to function as a living cell. Among these processes are the biochemical circuits that control cell division and differentiation. Four proteins serve as master regulators of these processes, Shapiro and her colleagues have found. Rising and falling quantities of these proteins in particular parts of the cell produce "an exquisite coordination of events in a three-dimensional grid."

Building these circuit diagrams has allowed researchers to identify nodes that control cellular functions and are attractive targets for drugs designed to alter the functioning of cells. Research in Shapiro's lab, for example, has led to drug development projects for two new antibiotics and an antifungal agent.

Shapiro's lab members are about half biologists and half physicists and engineers. Each has had to learn the language of the others so that they can work together. "You put all these people together and amazing things happen," Shapiro says. "Now we understand in a completely different way how this bacterial cell works."

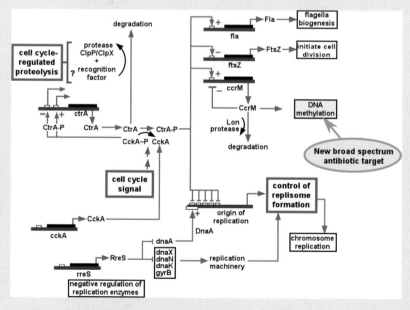

SOURCE: *Shapiro Lab, Stanford University School of Medicine.*

subdisciplines of biology to more problem-focused, collaborative and interdisciplinary research. Each is important, and many, if not most, biologists have feet in both worlds.

So if many biologists already practice the New Biology at some level, what is the role of this report? Its role is to bring attention to the remarkable depth and scope of the emerging New Biology that is as yet poorly recognized, inadequately supported, and delivering only a fraction of its potential.

Consider the newly hired assistant professor in the immunology department of a medical school who wants to collaborate with an ecologist who studies the impact of changing land use patterns on natural ecosystems and an engineer who models complex networks. Together they hope to develop a biosensor to detect emerging infectious diseases. Where will this group apply for funding? How will that assistant professor's tenure committee react to a series of publications in engineering and ecology journals?

Or consider the physics professor who wants to develop an interdisciplinary course on the physics and chemistry of DNA replication with colleagues from the chemistry and molecular biology departments. Will any of these professors be given credit for contributing to the teaching needs of their own departments? Such a course would likely not count toward degree requirements in any of the three departments. And yet the students who took such a course would be well-prepared to work across disciplinary boundaries no matter which of the three sciences they decided to pursue in depth.

Importantly, the New Biologist is not a scientist who knows a little bit about all disciplines, but a scientist with deep knowledge in one discipline and basic "fluency" in several. One implication of this is that not all "New Biologists" are now, or will in the future be, biologists! The physicists who study how the laws of physics play out in the crowded and decidedly non-equilibrium environment of the cell, or the mathematicians who derive new equations to describe the complex network interactions that characterize living systems are New Biologists as well as being physicists or mathematicians. In fact, the New Biology includes any scientist, mathematician, or engineer striving to apply his or her expertise to the understanding and application of living systems.

During its deliberations, the committee came to the conclusion that the best way for the United States to capitalize on the new capabilities emerging in the life sciences would be a multi-agency initiative to marshal the necessary resources and provide the coordination to enable the academic, public, and private sectors to address major societal challenges. The challenges laid out in this chapter are analogous to that of placing a man on the moon—the technologies do not all yet exist, there are still fundamental gaps in understanding—but the committee believes that a relatively small investment could reap enormous returns in each of these major societal challenge areas.

A NEW BIOLOGY APPROACH TO THE FOOD CHALLENGE: GENERATE FOOD PLANTS TO ADAPT AND GROW SUSTAINABLY IN CHANGING ENVIRONMENTS

The United Nations Food and Agriculture Organization has estimated that 923 million people were undernourished in 2007, an increase of 75 million over the 2003–2005 estimate of 848 million (FAO, 2008). Growing enough food worldwide to address this shortfall, as well as providing the higher quality food that will be expected by people living in countries where standards of living are improving, is an enormous challenge. This challenge will be compounded by the changing climatic conditions of the future, which will change the temperature and rainfall patterns of the world's farmlands, and may also lead to inundation of low-lying fertile land. A better fundamental understanding of plant growth and productivity, as well as of how plants can be conditioned or bred to tolerate extreme conditions and adapt to climate change, will be key components in increasing food production and nutrition in all areas of agriculture to meet the needs of 8.4 billion people by 2030 (Census Bureau, 2008), while allowing adequate land for energy production and environmental services.

Understanding Plant Growth

The long-term future of agriculture depends on a deeper understanding of plant growth. Growth—or development—is the path from the genetic instructions stored in the genome to a fully formed organism. Surprisingly little is now known about this path in plants. A genome sequence provides both a list of parts and a resource for plant breeding methods, but does not give the information needed to understand how each gene contributes to the formation and behavior of individual plant cells, how the cells collaborate and communicate to form tissues (such as the vascular system or the epidermis), and how the tissues function together to form the entire plant. There is simply a lack of fundamental information—we have the parts list for some plants, but not the assembly instructions, so we don't yet have a useful assembly manual. Understanding at a fundamental and detailed level of the assembly manual of even one plant would be a powerful tool. A recent NRC report, *Achievements of the National Plant Genome Initiative and New Horizons in Plant Biology* (National Research Council, 2008), provides a series of recommendations that could serve as the basis for planning a coordinated effort to understand plant growth, including a call to develop "reference genomes." The report details the benefits of such genomes and outlines the characteristics of desirable reference sequences. The NPGI report recognizes that sequencing is just a first step in understanding plant growth. A fully characterized model plant would provide a scaffold upon which the myriad variations found throughout the plant kingdom could be interpreted and put to use. Here lies the connection between biodiversity and meeting the challenge of revolutionizing our capacity to generate plant varieties to meet local needs and conditions. Every plant species, even every local population of a species, contains unique genetic resources that could contribute to improving the crops we depend on. Assessing and protecting biodiversity is therefore a critical ingredient in meeting the challenge of achieving sustainable food production.

Fundamental understanding will require predictive models that include all of the factors that affect growth and development; several technologies that will have to be developed to be applied in parallel to model and crop plants; new methods for live visualization of growing plants and for computational modeling of their growth and development at the molecular and cellular levels; cell-type specific gene expression, proteomic, and metabolomic data; high-throughput phenotyping, both visual and chemical; methods to characterize the dynamics and functions of microbial communities; and ready access to next-generation sequencing methods. The same technologies and measurement techniques will find useful application in energy, environmental, and human health research.

The New Biology—integrating life science research with physical science, engineering, computational science, and mathematics—will enable the development of models of plant growth in cellular and molecular detail. Such predictive models, combined with a comprehensive approach to cataloguing and appreciating plant biodiversity and the evolutionary relationships among plants, will allow scientific plant breeding of a new type, in which genetic changes can be targeted in a way that will predictably result in novel crops and crops adapted to their conditions of growth. The goal of predictability is a critical one; genuine understanding of plant growth will reduce uncertainty about any possible health or environmental consequences of genetic changes, changes in growth conditions, or in associated microbial or insect communities. The New Biology promises to deliver a dramatically more efficient approach to developing plant varieties that can be grown sustainably under local conditions. Advances in plant breeding and engineering, combined with a more profound and comprehensive understanding of plant growth and development and more complete knowledge of plant diversity, will make it faster and less expensive to develop plant varieties with helpful characteristics.

Genetically Informed Breeding

As a result of plant genome sequencing, plant genome analysis, and advances in bioinformatics, it is now possible to recast the principles of highly successful traditional plant breeding into a new and accelerated type of plant breeding termed "genetically informed breeding." Previously, the offspring of plant crosses had to be screened after their full life cycle to see which of them had acquired the traits sought by the breeder or farmer. Growing thousands of plant offspring required a lot of time and space, and therefore limited the numbers of offspring that could be analyzed. New quantitative methods—the methods of the

New Biology—are being developed that use next-generation DNA sequencing to identify the differences in the genomes of parental varieties, and to identify which genes of the parents are associated with particular desired traits through quantitative trait mapping. Once this is done, the genetic sequence, or geno-type, of millions of offspring can be determined from seeds or seedlings, and only those with the desired trait combinations retained. This will allow a much deeper selection from larger numbers of offspring, hence enormously speeding the overall rate and power of plant breeding.

Continued advances in genotyping (including the same next-generation sequencing methods that are contributing to the revolution in medicine, and the same bioinformatics of genetic association studies used in human genomics) and application of novel engineering methods to automatically record the relevant traits of growing plants, will greatly accelerate the process of breeding plants with desired characteristics.

Transgenics and Genetic Engineering of Crops

The advancement of plant genomics will also allow us to engineer crops in another way. By adding genes to the crop DNA from species other than the crop plant of interest, we may be able to capitalize on all of the many molecular mechanisms that can contribute to high crop yields. For example, some plants use an alternate photosynthetic pathway (called C4) that increases carbon fixation in dry environments. If the higher C4 photosynthetic rates could be transferred to crops that normally use conventional C3 photosynthesis, it could increase photosynthetic rates in most of the world's food crops. Or, manipulating the effects and concentration of hormones could optimize not just growth, but also partitioning of the carbohydrates produced by photosynthesis into grains and other edible parts of the plant. Additional advanced genetic and molecular methods, including those in place and others now being explored, are leading to improvement in the nutritional value of crops, for example by changing the composition of soybean oil to reduce transfat concentrations (Fehr, 2007).

Biodiversity, Systematics, and Evolutionary Genomics

Research in biodiversity, enhanced by rapid advances in comparative and evolutionary biology, is a critical ingredient in expanding the range of options available for developing new food crops and improving current ones. Information technology, imaging, and high-throughput sequencing are a few of the technological advances that promise to drive rapid advances in understanding and managing biological diversity. Developing a comprehensive knowledge of plant diversity and greater understanding of evolutionary relationships is the functional equivalent of building a fully stocked parts warehouse with an inventory control system that quickly locates exactly the right part. Much of this potential is as yet unrealized because most species on Earth are yet unnamed, indeed undiscovered, and their precise evolutionary relationships are unknown. The field of systematics—the study of the diversity of life and the relationships among organisms—is undergoing a renaissance as a result of adding genomic and computational analysis to the many other ways organisms can be compared. The practical benefits of expanding knowledge in this area are enormous; tapping into the vast resources represented by biological diversity will contribute to adapting and improving crops for food and bioenergy, understanding ecosystem function, and finding new biologically active chemicals for medical and industrial applications (Chivian & Bernstein, 2008).

Crops as Ecosystems

All crops grow in a complex environment, characterized by physical parameters like temperature, moisture, and light, and biological parameters including the viruses, bacteria, fungi, insects, birds, and others that interact with the crop plants. Therefore, greater understanding of insect-plant interactions, both beneficial and harmful, offers another route toward increasing crop productivity. Furthermore, complex microbial communities in the soil, previously difficult to study, play critical roles in providing nutrients and protecting plants from pests and diseases. Understanding these microbial communities in predictive detail will also point to new ways to increase plant productivity. Genetic engineering (as well as plant breeding) has been of great importance in improving crop resistance to plant diseases caused by viruses, bacteria, and fungi, and in resistance to herbivores such as insects.

Detailed understanding of how plants grow, a comprehensive catalogue of plant diversity and evolutionary relationships, and a systems approach to understanding how plants interact with the microbes and insects in their environments—each of these areas is ripe for major advances in fundamental understanding and none of them can be addressed by any one community of scientists. Molecular and cellular biologists, ecologists, evolutionary biologists, and computational and physical scientists will all be needed. Biomedical researchers with expertise in identifying and growing stem cells, neuroscientists with expertise in how neural networks monitor internal processes and seek out and respond to external signals, environmental engineers with expertise in monitoring and remediating contaminated ecosystems, hydrologists, soil scientists and meteorologists who study the physical systems that affect plant growth, private sector researchers with expertise in identifying promising research results and translating them into products—all of these and many others could make critical contributions, if their efforts can be coordinated.

The result of this focused and integrated effort will be a body of knowledge, new tools, technologies, and approaches that will make it possible to adapt all sorts of crop plants for efficient production under different conditions, a critical contribution toward making it possible to feed people around the world with abundant, healthful food, adapted to grow efficiently in many different, ever-changing local environments. At the same time, the tools and approaches developed through this effort will enhance the productivity of individual scientists around the world, whatever plant or ecosystem they are studying.

A NEW BIOLOGY APPROACH TO THE ENVIRONMENT CHALLENGE: UNDERSTAND AND SUSTAIN ECOSYSTEM FUNCTION AND BIODIVERSITY IN THE FACE OF RAPID CHANGE

Humans do not exist independently of the rest of the living world. From the most basic requirements of oxygen, clean water, and food, to raw materials like fuel, building material, fiber for clothing, and shelter that have allowed human societies to flourish around the globe, to intangible benefits that enrich the quality of life such as the shade of a tree on a hot day or the inspiration of an eagle in flight, humans are dependent on other organisms. Together, the resources and benefits that are provided by the living world are considered "ecosystem services" (Millennium Ecosystem Assessment, 2005). The amount of services that ecosystems can provide depends, at base, on their productivity: that is, their ability to use energy from the sun to make complex carbon-containing molecules like sugars and starches. Sustaining ecosystems so that their productivity remains high even in the face of rapid climate change is essential to sustaining and enhancing the quality of life of a growing human population.

Fundamental advances in knowledge and a new generation of tools and technologies are needed to understand how ecosystems function, measure ecosystem services, allow restoration of damaged ecosystems, and minimize harmful impacts of human activities and climate change. What is needed is the New Biology, combining the knowledge base of ecology with those of organismal biology, evolutionary and comparative biology, climatology, hydrology, soil science, and environmental, civil, and systems engineering, through the unifying languages of mathematics, modeling, and computational science. This integration has the potential to generate breakthroughs in our ability to monitor ecosystem function, identify ecosystems at risk, and develop effective interventions to protect and restore ecosystem function.

Monitor Ecosystem Services

Ecosystem services are varied and some of them are easier to measure than others. The amount of wood in a forest, for example, is easier to measure than the amount of protection a mangrove swamp will provide from coastal flooding. Measuring qualities like the impact of ecosystems on air and water quality, or placing a value on the carbon that is stored in undisturbed ecosystems, is challenging. Detecting changes in biodiversity and predicting the impact of extinctions on ecosystem services is even harder. But if the value of ecosystems is to be appreciated, the impact of human activities understood, and management decisions made on a scientific basis, it is important to develop a methodology and the necessary tools to monitor the state of ecosystems.

The New Biology has a great deal to offer in bringing together the necessary expertise and resources to implement a practical ecosystem monitoring system. Monitoring activities are already carried out by several agencies; the Environmental Protection Agency measures air and water quality, the National Science Foundation administers the National Ecological Observatory Network and Long Term Ecological Research Network programs, U.S. Geological Survey has the National Water Quality Assessment, the United States Forest Service carries out forest inventories, the Department of Energy and the National Aeronautical and Space Administration administer Ameriflux (which measures ecosystem level exchanges of CO_2, water, energy, and momentum across the Americas), and the Department of Agriculture carries out agricultural and soil inventories. Several nonprofit groups also maintain extensive databases of ecological information. However, each of these efforts measures different things, for different reasons, and the parts do not add up to a whole that provides the nation with a comprehensive understanding of the state of its ecosystems.

Ultimately, monitoring is required that is both intensive (covering ecosystem services in depth) and extensive (covering all kinds of ecosystems, and at regional and national scales). Current efforts focus on in-depth understanding of a few natural ecosystems and crisis intervention in damaged environments with measurements limited to a few key physical air or water quality characteristics.

Considerable research has already explored how ecosystem services can be measured. A 2000 NRC report, *Ecological Indicators for the Nation* (National Research Council, 2000a), recommended a set of national ecological indicators that would measure the extent and status of the nation's ecosystems, the nation's ecological capital, and ecological functioning or performance. The years since that report have seen many advances in ecosystem science, technology, and computational and mathematical approaches to describing ecosystem function. GIS, GPS, and remote sensing technologies (providing higher resolution and lowered costs) have led to rapid advances in ecology. GPS units are now cheap enough to be part of every ecologist's toolbox, allowing accurate mapping of species' distributions against existing maps of geological profiles, hydrological dynamics, and other environmental information. These technological advances have been as important in ecological research as inexpensive high-throughput sequencing has been in molecular, cell, comparative, and evolutionary biology.

The Heinz Center for Science, Economics, and the Environment, established in 1995, pulls together information from all of these sources to produce reports on indicators including two editions, in 2002 and 2008, of the comprehensive *State of the Nation's Ecosystems* (H. John Heinz III Center for Science, Economics, and the Environment, 2002, 2008). The Heinz Center acknowledges, however, that the current system is fragmented. A recent article by three Heinz Center directors (O'Malley et al., 2009) contended that—

> [A] coherent and well-targeted environmental monitoring system will not appear without concerted action at the national level. The nation's environmental monitoring efforts grew up in specific agencies to meet specific program needs, and a combination of lack of funding for integration, fragmented decision-making, and institutional inertia cry out for a more strategic and effective approach. Without integrated environmental information, policymakers lack a broad view of how the environment is changing and risk wasting taxpayer dollars.

The article goes on to point out that—

> [T]here are data gaps for many geographic areas, important ecological endpoints, and contentious management challenges as well as mismatched datasets that make it difficult to detect trends over time or to make comparisons across geographic scales. . . . In *The State of the Nation's Ecosystems 2008*, only a third of the indicators could be reported with all of the needed data, another third had only partial data, and the remaining 40 indicators were left blank, largely because there were not enough data to present a big-picture view.

No single scientific community, federal agency, or foundation can develop and implement a comprehensive set of ecosystem indicators, capable of monitoring the ecosystem services on which the nation relies. The New Biology approach—coordinating the resources already available and supporting research that integrates biology with physical and earth sciences, engineering, and computation—can be applied to build on such existing resources as the 2000 NRC report and the Heinz Center's 2002 and 2008 reports to evaluate potential ecological indicators in light of current capabilities and develop an implementable system for monitoring ecosystem function.

The goal of a monitoring system is to provide an accurate assessment of the services provided by ecosystems and to indicate when those services are at risk. The next step for the New Biology is to develop the knowledge and means to respond to the information provided by the monitoring system—to minimize the impact of human activities on ecosystem services and, even more importantly, to restore ecosystem function where it has been compromised.

Advance Understanding of Ecosystem Restoration

Medical doctors follow diagnosis with treatment options. The "medicine" of ecosystem treatment, however, has few arrows in its quiver. We do not currently have the tools needed to manage the biosphere. Between the two extremes of, on the one hand, preserving some ecosystems in their pristine state and, on the other hand, carrying out human activities with minimal regard to measuring or predicting their ecological impact, there are few options. The capacity to evaluate human impacts on ecosystem services and to provide options for minimizing or healing those impacts is another potentially valuable contribution of the New Biology.

A growing subfield in ecology is restoration, which ultimately holds the key to recovery of ecosystem services in heavily degraded areas (e.g., recovery of watershed function), and perhaps even to mitigation of climate change through designing ecosystems with even greater capacity for removing carbon from the atmosphere.[1] Ecological restoration has a role to play in improving crop productivity, reducing energy needs and slowing the loss of biodiversity. The question the New Biology can address is: Once we know that a system is at risk, how do we return it to a state that is more capable of providing ecosystem services?

Integrate Basic Knowledge about Ecosystem Function with Problem-Solving Techniques

Developing a science of ecosystem restoration will require integration of many fields of knowledge. For example, ecologists rely on soils science and hydrological studies for meaningful ecological niche modeling, which is heavily used in conservation (e.g., reserve design) and in climate change impact studies. Interdisciplinary work is already common as many ecosystem biologists reside in Earth Sciences Departments and institutes, and climate change biologists collaborate as often with meteorologists as with other biologists. Facilitating these efforts and integrating organismal, agricultural, evolutionary, and comparative biologists, engineers, computational scientists, and others to focus on the question of ecosystem restoration has the potential to provide treatment options for critical ecosystems. The New Biology could contribute to the development of a field one might call ecosystem engineering, analogous to the MD-PhDs of the biomedical field, grounded in both research and treatment.

A NEW BIOLOGY APPROACH TO THE ENERGY CHALLENGE: EXPAND SUSTAINABLE ALTERNATIVES TO FOSSIL FUELS

World annual requirements for energy grow at about the same rate as Gross Domestic Product (GDP) and are expected to increase by around 60 percent by 2030. Most of this increase will come from rapidly developing economies like India and China (IEA, 2008). More than three-fourths of the current need is currently met by fossil sources (EIA, 2007). The United States is no exception

---

[1] The July 31, 2009, issue of *Science* was devoted to this topic.

to this pattern, with our high dependence on such carbon-rich fossil fuels as oil, coal, and natural gas for our energy needs. Fossil fuels increase carbon dioxide emissions, which are linked to increased risk of global warming (Houghton & Intergovernmental Panel on Climate Change. Working Group I., 2001). A growing population, desiring a higher standard of living, has put even greater demands on our energy supplies. Thus we face the consequences of burning of high sulfur coal, depletion of petroleum reserves for transportation, and excess CO_2 being produced as stored hydrocarbon reserves are depleted. The environment is damaged both by the extraction of these resources and then by the subsequent release of the byproducts of their use. Sustainable, efficient, and clean sources of energy are crucial to reducing our dependence on and the depletion of fossil fuels. The New Biology can help propel the sustainable production of biofuels, and the United States could be the leader in this increasingly important industrial sector.

Direct conversion of biomass to thermal energy via combustion was our first source of energy. Improvements in biomass combustion continue, as does development of liquid fuels derived from thermochemical conversion of raw cellulose to liquid fuels. The major motivation for producing more biofuels is to reduce dependence on petroleum-based transportation fuel. In 2007, Congress passed the Energy Independence and Security Act (Public Law 110-140). Among other things, the legislation included the Renewable Fuel Standard (RFS) program, which calls for the volume of renewable fuel required to be blended into gasoline from 9 billion gallons in 2008 to 36 billion gallons by 2022. In 2007, the United States consumed 176 billion gallons of fuel for transportation, so 36 billion gallons of renewable fuel (assuming equivalent energy content per unit volume) would cover roughly 20 percent of our transportation fuel needs (Bureau of Transportation Statistics, 2009). The RFS further stipulates that a substantial fraction of the biofuel must be advanced or cellulose-based fuels, rather than ethanol derived from corn.

In fact, technology is not currently available to meet the RFS, but the New Biology offers the possibility of advancing the fundamental knowledge, tools and technology needed to achieve it. Making efficient use of plant materials—biomass—to make biofuels is a systems challenge, and this is where the New Biology can make a critical contribution. At its simplest, the system consists of a plant that serves as the source of cellulose and an industrial process that turns the cellulose into a useful product. There are many points in the system that can be optimized: choosing the right crops as sources of biomass, engineering these crops so that they grow with a minimum of energy, fertilizer, and water input and produce cellulose that is easy to ferment, and engineering enzymes that are efficient at digesting the cellulose. The optimization of each of these steps depends on the others, so maintaining a view of the whole system is important.

Each of these steps involves a large number of choices. Which plants can produce the most biomass with the least input of fertilizers and water and the least impact on the land needed to grow food and sustain ecosystem services? How can those plants be modified to produce biomass even more efficiently or produce cellulose that is easier to digest? How can that biomass be converted to fuel? What enzymes and metabolic pathways have evolved in microbes that could be adapted for biomass conversion? How can the fermentation process be optimized to produce the most fuel at lowest cost? Work in all of these areas is underway in laboratories and companies, but the New Biology approach can tackle this challenge in a comprehensive way, bringing together expertise from many different scientific communities, federal agencies, and the private sector to generate the advances in fundamental knowledge and translational and developmental research needed to provide innovative solutions.

Identifying and Optimizing Sources of Biomass for Biofuel

Today, the United States leads the world in the volume of biofuels produced, and nearly all U.S. biofuel today is made by using fermentation to convert starch from corn into ethanol. More than 30 percent of the U.S. corn crop is used for ethanol production (USDA, 2009). As a result of the application of biotechnology to agriculture, per-acre corn yields are increasing at 2 to 3 percent per year (Egli, 2008). Grain alone, however, will not allow dramatic expansion of biofuel production, and must be supplemented, and ultimately replaced, with other sources of biomass. Development of energy crops that are direct sources of fermentable sugars, such as sugarcane or sweet sorghum, or sources of cellulosic materials, such as switchgrass, miscanthus, or agricultural and forestry byproducts, is an important priority. The same fundamental knowledge, tools, and technologies developed in the New Biology approach to the food challenge would be directly applicable here: understanding plant growth; advancing genetically informed breeding, transgenics and genetic engineering; advancing biodiversity, systematics and evolutionary genomics; and understanding crops as ecosystems. Thus, both the agriculture and energy research communities will be stakeholders in the effort to transform plant breeding capabilities.

Identifying and Optimizing Microbial Biocatalysts

Ethanol is a first-generation biofuel, produced via fermentation of sugars by wild-type yeast. Ethanol has limitations, such as low energy density, high vapor pressure, and water solubility. By combining recent advances in technologies such as high-throughput sequencing, automated gene expression measurement, and metabolic engineering, future generations of biofuels are now within reach. Advanced biofuels targets include higher alcohols, long-chain fatty acids and derivatives, even olefinic and alkane derivatives—all products that can be made by microorganisms, with subsequent chemical processing in some cases.

For each fuel molecule, microbial hosts are being selected, metabolic pathways are being identified, and using recombinant methods, organisms are being engineered to deliver biofuels at acceptable rates and yields. The challenge for advanced biofuels is to be able to produce fuel more cheaply than using yeast to ferment starch or sugar into ethanol. Again, the unique contribution the New Biology can add to these existing efforts is the coordination of efforts to discover, characterize, and engineer microbes so that they serve as factories for high production rates, with efforts to engineer production systems that maximize those microbes' productivity, for example by continuously adjusting levels of nutrients and end-products. These optimized systems will allow next-generation biofuels to compete with gasoline at prevailing prices.

Approaching Biofuel Production as a Systems Challenge

Clearly, the road to meeting the U.S. Renewable Fuel Standard consists of multiple steps, steps that are interdependent. An integrated approach that includes scientists and engineers expert at each step is essential. The combined efforts of plant scientists, microbiologists, ecologists, chemical and industrial process engineers, molecular biologists, geneticists, and many others are needed to develop and optimize the biomass-to-biofuel system. Combining the strengths of these communities does not necessarily mean bringing these experts into the same facility. Indeed, advancing communication and informatics infrastructures make it easier than ever to assemble a virtual collaboration. The New Biology Initiative proposed in this report would provide the resources to attract the best minds from across the scientific landscape to the problem, ensure that innovations and advances are swiftly communicated, and provide the tools and technologies needed to succeed.

A coordinated effort to optimize the conversion of biomass to biofuel would create knowledge and technologies that would have an immediate and direct impact on other sectors, including therapeutics and industrial materials, which can also be produced in this way (Box 2.2).

A NEW BIOLOGY APPROACH TO THE HEALTH CHALLENGE: UNDERSTANDING INDIVIDUAL HEALTH

The New Biology approach to the environmental challenge aims to make it possible to monitor ecosystem function and restore that function when it is compromised. The goal of a New Biology approach to health is similar—to make it possible to monitor each individual's health and treat any malfunction in a manner that is tailored to that individual. In other words, the goal is to provide individually predictive surveillance and care. In both cases, reaching these goals means understanding how the interactions of myriad components are related to overall system function.

BOX 2.2
Biomass to Biomaterials and Biosynthesis of
Chemicals and Therapeutics

Liquid transportation fuels represent by far the largest-volume opportunity for renewable products made by living organisms from biomass. Chemicals and materials made from petroleum account for about 10 percent of total U.S. oil consumption. The same fermentation-based technology that is being developed for biofuels can also be harnessed to replace petroleum-based materials and make many other useful products.

Microbes can also be engineered to produce chemicals, industrial enzymes, and therapeutics at industrial scales. Insulin was the first material produced using an engineered organism, in the 1980s. Many pharmaceutically active proteins, antibiotics, vitamins, and amino acids for food and animal feed followed. Microorganisms are now widely used to produce industrial enzymes for many uses. Fermentation-based large-scale (>10 ktons/year) production of chemicals and materials is an emerging opportunity. For example, lactic acid, produced by bacteria, can be polymerized into a substance that has useful properties in both fiber and molded form. Large-volume production of propanediol (PDO), using an engineered *E. coli* host, was commercialized in 2007. PDO is used in applications ranging from deicing fluids to personal care products. BioPDO is also an important component of poly (trimethylene terephthalate), a polymer with true engineering properties. We can now foresee a range of such chemical building blocks as succinic acid, dodecanedioic acid, and p-hydroxybenzoic acid being made from biomass via fermentation (Pacific Northwest National Laboratory, 2009).

At present, medical decision-making is often based on probabilities. For example, high cholesterol levels are associated with heart disease and early-stage cancers metastasize at a predictable rate. But some individuals with high cholesterol do not develop heart disease, and metastasis of a given tumor type occurs with frightening speed in some individuals and not at all in others. Each individual has a unique set of genes and a unique environmental history, yet the relationship of all of this variation to health is uncertain. Understanding the relationship of an individual's genetic makeup and environmental history to that individual's health risks, disease susceptibility, and response to treatment is a challenge well beyond current capabilities. Critical to improving that understanding is a quantum leap in our ability to understand the functioning of and interactions among complex networks, or systems of interconnected components.

The Genotype-Phenotype Challenge

It seems likely that it will soon be economically feasible to determine the full genome sequence of every individual. An individual's genetic make-up, or genotype, is directly related to his or her phenotype (i.e., the various traits of an individual that can be observed or measured). Because genetic sequences serve as the blueprints for all biological processes, genetic variation affects the functioning of all of the networks that underlie human health.

As if the challenge of understanding the connection between an individual's genome sequence and health were not difficult enough, two additional factors add further layers of complexity. First, feedback from the environment affects how the genetic blueprint is executed. For example, individuals who live at high altitudes, where the air holds less oxygen, produce more red blood cells. All individuals have the genetic potential for this adaptation, but it only occurs under particular environmental circumstances. Diet, exercise, exposure to sunlight, chemicals, viruses, and bacteria—all of these and much more can affect the connection between genotype and phenotype. Furthermore, new kinds of gene regulation continue to be discovered, including epigenetic mechanisms (mechanisms that change gene expression without changing the underlying gene sequence) and mechanisms like small, interfering RNA, in which short RNA fragments regulate expression or translation.

The second layer of complication recently added to the challenge of understanding the genotype-phenotype connection is the discovery that our own human genes are not the only genetic material affecting our health. Humans are intimately associated with a complex microbial community—the microbiome. Rapidly accumulating discoveries of the many essential roles of this microbial consortium are redefining our understanding of human health and making it clear that a true understanding of human health must take into account not only the human genome, but also the genomes of each human's microbial community. For example, differences in the microbiomes of twins has been shown to be associated with obese versus lean physiological states (Turnbaugh et al., 2009). The normal, healthy human body contains ten times more microbes than human cells, and these microbes carry out many essential functions. The microbes in the human intestine synthesize essential amino acids and vitamins, and digest complex carbohydrates (Backhed et al., 2005). Genomic and other new technologies are now making it possible for life scientists to characterize the human microbiome and the factors that influence the distribution, function, and evolution of our microbial partners. Recognizing the importance of the microbiome means assessing how these evolutionarily ancient microbial partnerships influence health and predisposition to diseases. In other words, the connection of genotype to phenotype must include not only the human genome, but also the genomes of the microbes that live in and on us. The important influence of viruses on the genotype-phenotype connection must also be taken into account, from their role in cancer (for example, HIV and Kaposi's sarcoma, human papilloma virus and cervical cancer) to the role that the immune response to persistent viruses plays in the development of autoimmune and other chronic diseases. Understanding the role of microbes and viruses in human health is a major challenge, but it also holds the promise of providing new intervention points for prevention, diagnosis, and treatment of disease.

A growing body of evidence suggests that many diseases, including types 1 and 2 diabetes, coronary artery disease, and glioblastoma, typically result from small defects in many genes, rather than catastrophic defects in a few genes (Altshuler et al., 2008). It is likely that many different combinations of genetic changes, acting in the context of particular environmental influences (for example, a viral infection), can produce the same disease, so that understanding how genes work together in regulatory networks, and how those networks are affected by external factors, will be crucial to untangling the intricate web of interactions associated with a particular disease phenotype.

Large-scale studies that associate genotype to phenotype are rapidly identifying many, many genetic variations (both human and microbial) and environmental factors that are associated with specific diseases. The key word in that sentence is "associated." While some of these variations may have a direct role in causing disease, there is currently a substantial gap between discovering an association and uncovering a causal mechanism. But ultimately, if health care is to move from treatment based on statistical likelihood to treatment based on each individual's specific circumstances—in other words, truly personalized medicine—the chasm between genotype and phenotype will have to be bridged. This is a challenge that is beyond the scope of any single Institute at the NIH. Indeed, it is a challenge that will demand a New Biology-driven research community empowered by scientific and technical resources from across the federal government, the broad community of scientists, and the private sector. Unraveling the genotype-phenotype connection will require combining increasingly sophisticated genotype-phenotype associations with experimentation, modeling, systems analyses, and comparative biology.

Understanding Networks

Between the starting point of an individual's gene sequences and the end-point of that individual's health is a web of interacting networks of staggering complexity. Recent advances are enabling biomedical researchers to begin to study humans more comprehensively, as individuals whose health is determined by the interactions between these complex structural and metabolic networks. On the path from genotype to phenotype, each network is interlocked with many others through intricate interfaces, such as feedback loops. Study of the complex networks that monitor, report, and react to changes in human health is an area of biology that is poised for exponential development. These networks consist of circuits of interacting genes, gene products, metabolites, and signals that function together much like electronic integrated circuits. Unlike electronic circuits, however, almost all of the components of living networks are constantly changing, with results that ripple through all of the other networks. Tools and methodologies are being developed that can detect, synthesize, and process complex biological information at a network level, image cellular events in real time, delineate how proteins interact, and access single sites within the DNA library of the cell. Computational and modeling approaches are beginning to allow analysis of these complex systems, with the ultimate goal of predicting how variations in individual components affect the function of the overall system. Many of the pieces are identified, and some circuits and interactions have been described, but true understanding is still well beyond reach. Combining fundamental knowledge with physical and computational analysis, modeling and engineering, in other words, the New Biology approach, is going to be the only way to bring understanding of these complex networks to a useful level of predictability.

Such complex events as how embryos develop or how cells of the immune system differentiate (that is, the actual processes by which an individual's phenotype—the appearance and characteristics of the individual—come into being) must be viewed from a global yet detailed perspective because they are composed of a collection of molecular mechanisms that include junctions that interconnect vast networks of genes. It is essential to take a broader view and analyze entire gene regulatory networks, and the circuitry of events underlying complex biological systems. Data obtained from mutational, chemical genetic or imaging analyses of organisms such as *Drosophila, C. elegans, Arabidopsis*, mouse, sea urchin, and other species will continue to uncover rich sets of interactions between gene products that comprise such regulatory networks. Analysis of developing and differentiating systems at a network level will be critical for understanding complex events of how tissues and organs are assembled. These studies have obvious import in regenerative medicine.

Similarly, networks of proteins interact at a biochemical level to form complex metabolic machines that produce distinct cellular products. Understanding these and other complex networks from a holistic perspective offers the possibility of diagnosing human diseases that arise from subtle changes in network components.

Perhaps the most complex, fascinating, and least understood networks involve circuits of nerve cells that act in a coordinated fashion to produce learning, memory, movement, and cognition. These studies can be approached both experimentally, at the cellular level, as well as at the whole brain level via functional imaging approaches. In addition, recent computational approaches allow modeling of neurobiological systems that provide valuable predictive information.

Understanding networks will require increasingly sophisticated, quantitative technologies to measure intermediates and output, which in turn may demand conceptual and technical advances in mathematical and computational approaches to the study of networks.

Studying Complex Systems Directly in Humans

Work in model organisms, as discussed earlier, is immensely productive because fundamental developmental and metabolic pathways have been conserved throughout evolution and are shared among many organisms. In fact, model organisms are increasingly useful as genomics makes it possible to understand the differences and similarities among organisms at an ever more detailed level. Advances in imaging, high-throughput technologies, and computational biology increasingly make it possible to relate model system information directly to the study of complex systems in the human. New technologies and sciences that allow, for example, comprehensive comparisons of genomes and gene expression will enable much more sophisticated associations between genotype and phenotype.

Another approach to the genotype-phenotype challenge is to survey the "read-out" from the genome: that is, the collection of proteins and metabolites that are the end-products of gene activities. Technologies for characterizing proteomes (all the proteins in a sample) and metabolomes (all the metabolites in a sample) are less capable and more expensive than sequencing technologies. But they are increasingly being used to generate profiles from body fluids, such as blood, sweat, and urine, which contain the products and byproducts of metabolic processes that reflect a composite of an individual's genomic activity together with that of his or her particular microbiome. These profiles can be used, for example, to design tailor-made drugs (i.e., drugs that would take into account differences in how individuals break down and assimilate a given pharmaceutical). Ultimately, high-throughput assessment of the metabolome could provide a remarkably precise picture of the overall activities within and on the human body and critical insights into the relationship of "composite genotype" to phenotype. Achieving those insights, however, will require new technologies to measure proteins and metabolites, massive collections of samples from healthy and sick individuals, and novel mathematical and computational tools, and concepts to discern the patterns associated with health and disease. Such a task dwarfs the complexity of the Human Genome Project.

A Systems Approach to the Genotype-Phenotype Challenge

Unraveling the genotype-phenotype connection will require that the efforts of biomedical researchers be complemented and supplemented by the skills and different approaches of engineers, mathematicians, and physical and computational scientists. The efforts of scientists nurtured by separate institutes of the NIH will need to be joined by those supported by NSF and DOE, for example, who study non-human organisms and who create and support various multi-user facilities. Agencies like the National Institute of Standards and Technology could add interdisciplinary expertise in the development of measurement technologies, pharmaceutical companies their extensive databases, and nonprofit disease research foundations their refined expertise. By providing the framework for these communities to work together to address the challenge of understanding the genotype-phenotype connection, the New Biology can accelerate fundamental understanding of the systems that underlie health and the development of the tools and technologies that will in turn lead to more efficient approaches to developing therapeutics (Box 2.3). And just as with sequencing technology, the technological and conceptual breakthroughs that emerge from these efforts could revolutionize the capacity and sophistication of all biological research.

INTERCONNECTED PROBLEMS, INTERCONNECTED SOLUTIONS

The future holds truly imposing challenges for humankind: efficiently improving the sustainable productivity of diverse food crops, producing sustainable substitutes for fossil fuels, monitoring and restoring ecosystem services, and understanding and promoting human health. The New Biology described in this report, if properly nurtured and supported, has the potential to contribute to real progress in meeting these challenges and many tools and approaches will be shared for all four problem areas. The projected impacts are significant, from both a societal and economic perspective. Furthermore, the importance of the challenges to which the New Biology will contribute ensure

---

BOX 2.3
Developing Therapeutics to
Prevent, Treat, and Cure Disease

The future of therapeutics lies in the application of new technologies as tools for detecting and treating diseases. Therapeutic efforts will also benefit from an increased understanding of networks. Therapeutics that focus on a single driver may miss both the critical role played by other genes as well as the ease with which a malignant cell, for instance, may utilize alternative parts of the larger network to side-step the drug's effect, and thus continue to thrive. Similarly, adverse side effects can result when intervention in one network causes unforeseen changes in others. Complicated as these networks are, we are now in a position to study the response of complex systems to a range of perturbagens (both natural mutations and introduced chemicals), providing an important opportunity to probe the pattern of interactions and refine the model. This approach may also identify underappreciated network pressure points—possible drug targets or biomarkers that are less evident in traditional linear models.

that students, and the American public, will be inspired to help, and will be drawn in to science and science education. The United States cannot afford to wait for others to create these life science-based solutions. As a nation, we must lead these efforts.

3

Why Now?

The moment is ripe to invest in the development of the new biology because the life sciences are in the midst of a historical period analogous to the early 20th century in the physical sciences. The discovery of the electron in 1897 marked the beginning of a major turning point in the history of science. Over the next few decades, physics, chemistry, and astronomy were all transformed. Physicists uncovered the fundamental constituents of matter and energy and discovered that these constituents interact in unanticipated ways. Chemists related the structure and properties of substances to the interactions of electrons surrounding atomic nuclei. Astronomers related the light received from stars and the sun to the chemical properties of the atoms generating that light. In this way, new connections within the physical sciences became apparent and drove further advances.

These theoretical advances also led to practical applications that transformed society. Having a "parts list" of the physical world enabled scientists and engineers to develop technologies that would not have been possible without this understanding. These technologies led in turn to the electronic industry, the computer industry, and the information technology industry, which together have created a world that could scarcely have been imagined a century ago.

Before the transition associated with the discovery of the electron, scientists gathered increasing amounts of data, but those data could not be put to full use because of the lack of a conceptual framework. After that discovery, previously gathered data took on new usefulness, entirely new areas of inquiry emerged, and discovery and application accelerated rapidly. Such discoveries are critical junctures that send science and society in new directions.

These moments of rapid acceleration of scientific progress have different origins. Some occur due to technological advances, such as the invention of the telescope or microscope. Some occur due to conceptual advances, such as the description of evolution by natural selection or the development of relativity theory. In some cases there is a principal driver; in others, multiple factors combine to accelerate progress. New discoveries and new technologies do not guarantee that discovery will accelerate. The world must be ready for change, and the tools and resources must be available to capitalize on new capabilities or knowledge.

This committee believes that the life sciences currently stand at such a point of inflection. Drawing ever nearer is the possibility of understanding how all of the parts of living systems operate together in biological organisms and ecosystems. This understanding could have a profound influence on the future of the human species. It could help produce enough food for a growing population, prevent and cure chronic and acute diseases, meet future needs for energy, and manage the preservation of Earth's biological heritage for future generations.

The approach of this moment of opportunity in the life sciences has become increasingly evident over the last decade, and in many ways, the New Biology has already begun to emerge. It has become common to hear statements that the 21st century will be the century of biology. What are the factors that have brought biology to this point? And what current ideas, tools and approaches represent the emergence of new capabilities?

THE FUNDAMENTAL UNITY OF BIOLOGY HAS NEVER BEEN CLEARER OR MORE APPLICABLE

The great potential of the life sciences to contribute simultaneously to so many areas of societal need rests on the fact that biology, like physics and chemistry, relies on a small number of core organizational principles. The reality of these core commonalities, conserved throughout evolution—that DNA is the chemical of inheritance, that the cell is the smallest independent unit of life, that cells can be organized into complex, multicellular organisms, that all organisms function within interdependent communities and that photo-systems capture the solar radiation to provide energy for all of life processes—means that any knowledge gained about one genome, cell, organism, community, or ecosystem is useful in understanding many others. Because living systems are so complex, much biological experimentation has had to focus on individual or small numbers of components within a single organizational level. The reductionist approach has helped reveal many of the basic molecular, cellular, physiological, and ecological processes that govern life.

This work needs to continue in the future. Many aspects of biological function remain unknown on all levels. But biologists are now gaining the capability to go beyond the interactions of components within a single level of biological organization and the study of one or a few components at a time. As microbiologist and evolutionist Carl Woese said over 30 years ago, "Our task now is to resynthesize biology; put the organism back into its environment; connect it again to its evolutionary past. . . . The time has come for biology to enter the nonlinear world (Woese & Fox, 1977). The practical ability to achieve Woese's vision is now beginning to emerge. Biologists are increasingly able to integrate information across many organisms, from multiple levels of organization (such as cells, organisms, and populations) and about entire systems (such as all the genes in a genome or all the cells in a body) to gain a new integrated understanding that incorporates more and more of the complexity that characterizes biological systems (Box 3.1).

As the biological sciences advanced during the 20th century, separate fields emerged to tackle the complex subsystems that together make up living systems. Genetics, cell biology, ecology, microbiology, biochemistry, and molecular biology each took on various aspects of the challenge. The sheer volume of knowledge generated in each of these subdisciplines made it increasingly difficult for researchers who studied organisms to keep up with the progress being made by researchers studying cells, and those studying molecules rarely interacted with those studying ecosystems. Scientists in each of these specialties

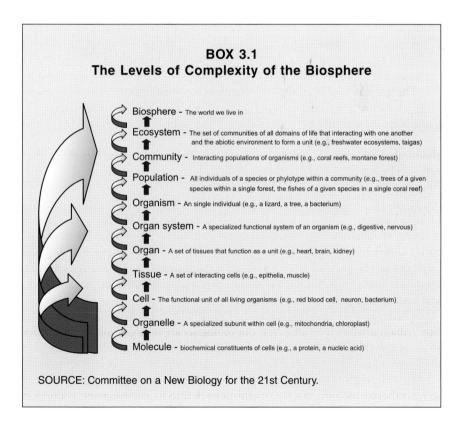

BOX 3.1
The Levels of Complexity of the Biosphere

Biosphere - The world we live in

Ecosystem - The set of communities of all domains of life that interacting with one another and the abiotic environment to form a unit (e.g., freshwater ecosystems, taigas)

Community - Interacting populations of organisms (e.g., coral reefs, montane forest)

Population - All individuals of a species or phylotype within a community (e.g., trees of a given species within a single forest, the fishes of a given species in a single coral reef)

Organism - An single individual (e.g., a lizard, a tree, a bacterium)

Organ system - A specialized functional system of an organism (e.g., digestive, nervous)

Organ - A set of tissues that function as a unit (e.g., heart, brain, kidney)

Tissue - A set of interacting cells (e.g., epithelia, muscle)

Cell - The functional unit of all living organisms (e.g., red blood cell, neuron, bacterium)

Organelle - A specialized subunit within cell (e.g., mitochondria, chloroplast)

Molecule - biochemical constituents of cells (e.g., a protein, a nucleic acid)

SOURCE: Committee on a New Biology for the 21st Century.

attended separate meetings, published in different journals, and generally had little communication. Although it was understood on a conceptual level that the organizational levels are tightly interlocked, most researchers focused on a single system in great detail.

Recently, though, the connections among the fields of the life sciences have become easier to study. For example, tools and concepts that arose within individual subdisciplines within the life sciences are now applied throughout biology. Thus, biochemistry and molecular biology are now techniques that are applied nearly universally across the life sciences. Genomic data and techniques have widespread applications in biology and reveal the connections among fields. In particular, genomic comparisons reveal the common descent of organisms and enable researchers to make comparisons of different types of organisms (Box 3.2), while also highlighting the differences that have arisen in separate evolutionary lineages. These cross-species investigations have started to blur the boundaries between such fields as microbiology, botany, and zoology. Discovering and understanding the features shared by all living organisms, and the differences that make each system or organism unique, has never been easier or more productive.

Despite the development of common tools, questions, and methodologies, scientists within subdisciplines that were historically separate still do not have the optimal level of interaction. This is particularly true when the goal is to develop new science linking multiple levels of organization in biological systems. To accelerate progress in the life sciences, researchers from different subdisciplines need to interact and collaborate to a greater extent. Presenting these communities with a common problem to solve will provide an opportunity for them to bring their different skills and perspectives to bear and accelerate the development of conceptual and technological approaches to understanding the connections between the different levels of biological organization (Box 3.3).

NEW PLAYERS ARE ENTERING THE FIELD, BRINGING NEW SKILLS AND IDEAS

Just as integration is becoming more important *within* the life sciences, immense value is emerging from the integration of the life sciences with other disciplines. For example, the recent and continuing revolution in genomics has come from an integration of techniques and concepts from engineering, robotics, computer science, mathematics, statistics, chemistry, and many other fields. The precipitous decline in the cost of genome sequencing would not have been possible without a combination of engineering of equipment, robotics for automation, and chemistry and biochemistry to make the sequencing accurate. Similarly, expertise from fields as diverse as evolutionary biology, computer science, mathematics, and statistics was necessary to analyze raw genomic data and to extend the use of these data to other fields.

BOX 3.2
Common Descent and the
Integration of the Life Sciences

Though every species on the planet is unique in some ways, all species are linked to each other by common descent: that is, any two species evolved from a common ancestor that lived at some point in the past. As a result, all species share some biological properties due to the inheritance of features present in their common ancestor. Work in one species can be of direct relevance to the understanding of other species because processes may be identical or highly similar between the two due to their shared descent. This is part of the reason for the importance of "model organisms." Studies in mammalian model systems such as the mouse have led to major insights into human biology. Examples include studies of cholesterol metabolism (which led to the development of the statins, a class of drugs that have dramatically reduced atherosclerosis and cardiovascular disease; beta adrenergic receptors, which led to the development of beta blockers for the treatment of hypertension and heart disease; and tumor necrosis factor, which led to the development of therapeutic antibodies that provide relief to people with rheumatoid arthritis.

Frequently, discoveries in one organism have implications even for very distantly related organisms. The degree of relatedness of two organisms, which is determined in large measure by the amount of time that has elapsed since their common ancestor, indicates how much of their biology they share. Consequently, the older a biological process, the more likely that it will be shared by a great number of organisms. Thus, an important mechanism for regulating protein levels in cells, called "small interfering RNA" or siRNA, was initially discovered in plants, but was then found to be at work in human cells and shows great promise as a new approach for drug development (Carthew & Sontheimer, 2009). Even more distantly related organisms can share common genes and pathways. Studies of mutational processes in the bacterium *Escherichia coli* and the yeast *Saccharomyces cerevisiae* helped identify the genes that are defective in hereditary nonpolyposis colon cancer (HNPCC) in humans (Fishel & Kolodner, 1995).

Because no two species are exactly the same, work in model organisms does not always translate perfectly into other species, even if the two species are quite closely related. For example, the most promising AIDS vaccine candidate failed in human clinical trials (Buchbinder et al., 2008) despite showing promise in experiments with monkeys (Shiver et al., 2002). Animal models of neurodegenerative disease, such as Alzheimer's or Parkinson's, and of psychiatric diseases, such as schizophrenia or depression, do not fully reproduce the clinical signs seen in humans with these disorders. Moreover, drugs shown to have efficacy in such animal models have failed in human clinical trials. Better understanding of which characteristics are shared and which are not is a major outstanding challenge in biology, which, when met, will greatly improve our ability to predict how results in one organism will apply to another.

BOX 3.3
Eco-Evo-Devo: Integration across Subdisciplines

The field of evolutionary-developmental biology (known as evo-devo) has emerged in the last 15 years. It offers a powerful example of the potential for integration of biological theory and practice across the hierarchy of life, from molecules to ecosystems. Studies in evo-devo have demonstrated that different animal body plans can result from the alternative expression patterns of a "toolbox" of conserved genes (such as the homeobox genes) and gene networks. The research frontiers of this field lie in the continued development of computational and mathematical tools for the study of the links between development and evolution, and determination of the environmental cues that underlie developmental processes over evolutionary time and within the life of a given individual.

As an indication of the kind of mathematical and computational tools needed, analyses of gene sequence data to derive the phylogeny of the animal kingdom required the full-time use of 120 processors over several months (Hejnol et al., in press). Widespread application of such approaches is prohibited by technical constraints, demonstrating the need for significantly more efficient means of computational analysis for such large datasets.

The field of evo-devo is now poised for integration across the entire hierarchy of molecular through organismal biology. For example, at the broadest levels, biologists have long known that the environmental parameters, such as temperature and length of daylight, can profoundly influence developmental processes. However, until recently, developmental biologists have focused almost entirely on an understanding of the gene-to-organism aspects of development—that is, they have studied how the cells and tissues of an organism progress from fertilization through death. Much must still be done to understand these processes, but at the same time they can now be placed within a context of the "ecology of development," or eco-devo.

Developmental processes are often the "canaries" of environmental perturbation. For example, the biogeographic distributions of many marine animals are determined by the temperature sensitivity of their larval stages. The distribution of large numbers of marine taxa is likely to be dramatically altered by either changes in ocean temperatures or in ocean circulation patterns. The ripple effects of such changes would affect many aspects of human interface with the oceans, including fisheries. Predicting and minimizing the impact of environmental changes on development, species ranges, and ecosystem services will require collaboration among developmental biologists, ecologists, computational scientists, oceanographers, climate scientists, and others to integrate their knowledge of different parts of the system.

The need to analyze these data has driven a rapid expansion in the application of mathematics to biology. In particular, two aspects of computation have been critical. First, algorithms and computational power for analysis of large data sets help make sense of the massive amounts of data produced by genomic studies. Second, the placement of data in accessible digital databases has greatly improved the ability to share information and build on prior work.

Such advances in computation have been critical in many areas of biology, such as ecosystem studies, conservation biology, evolutionary biology, and epidemiology. Each of these fields needs to handle large complex data sets, digitize and share information, and test complex models and theories. To carry out this work, biology has taken advantage of developments from other fields such as physics, astronomy, and earth sciences, which have been for many years handling and analyzing massive data sets. Biological data sets are especially challenging because they must cover so many diverse organisms and measure many different characteristics that are constantly changing. Connecting those data sets is extremely difficult but absolutely essential. Success will demand close cooperation between the biologists and other scientists who study the system, computer scientists and mathematicians who develop new ways to analyze the data, and engineers who bring expertise in modeling.

The issue of predictability is one major reason why even the present level of integration of life sciences with engineering is already productive. Engineering offers a way of thinking that can contribute substantively to unraveling the inherent complexity of biological science. The essence of engineering is predictive design. Engineers seek to create systems that can operate reliably within some circumscribed conditions. Moreover, engineering design is almost always undertaken in the face of incomplete information. Even with technologies based on the physical and chemical sciences, there remain many poorly characterized parameters. These limitations also apply to the biological systems, even under the best of circumstances, so bringing an engineering mindset to bear on biological questions is already beginning to add a new layer of value to basic biological research.

Already, large numbers of physicists are being drawn into the biological sciences and teams that include engineers, earth scientists, biologists, computational scientists, and others are beginning to conceive and approach biological research in new ways (Boxes 3.4 and 3.5). What needs to occur next is for the boundaries between disciplines to be broken down even further, much as the boundaries within biology are being broken down. Making the New Biology a reality will require not only the best in technology and science, but also a uniquely interdisciplinary approach. Efforts to date must be seen as a "first pass" rather than as a complete integration across multiple fields. The eventual goal is for all scientists and engineers who study biological systems, whether their in-depth training is in physics, mathematics, or chemical engineering, to see themselves also as New Biologists, together contributing to the emergence of the New Biology.

A STRONG FOUNDATION HAS ALREADY BEEN BUILT

Over the past 40 years, large investments have produced remarkable discoveries in the biological sciences. These discoveries have in a large part come

BOX 3.4
Brain-Machine Interfaces

Brain-machine interfaces are systems that allow people or animals to control an external device through their brain activity. In 2003, scientists demonstrated that monkeys with electrode implants in their brains, connected to a robotic arm, could manipulate a robotic arm using only their thoughts. In 2008, scientists demonstrated for the first time that brain signals from a monkey could make a robot walk on a treadmill. Scientists hope that such technology will be a great benefit for people who are paralyzed or no longer have control of their physical movements. Such technology and experiments will also lead to an increased understanding of how the brain works.

Brain-machine interfaces are an example of the convergence of different areas of science and technology, and the importance of encouraging the emergence of the New Biology as an integrated science. In the case of the brain-machine interface that permitted monkey thoughts to make a robot walk, the electrodes were placed in the part of the brain that earlier neurobiological studies had shown contained neurons that fired when primates walk. Detailed video images of leg movements were then combined with measurements of simultaneous brain cell activity, and then analyzed using sophisticated computational methods. A robot, previously designed to closely mimic human locomotion, was then programmed to respond to the brain signals in the monkey (Blakeslee, 2008).

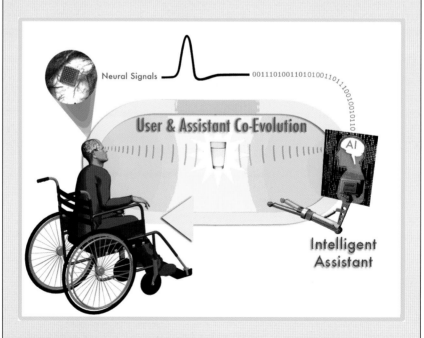

SOURCE: Sanchez et al., 2009.

BOX 3.5
Nanotechnology—The Artificial Retina

The intertwined nature of the physical and life sciences is exemplified in the progress that has been made with the artificial retina, a device that resulted from a multi-laboratory initiative supported by the Department of Energy. The device has already shown promise in patients with macular degeneration, a major cause of blindness in the elderly. The nerves that are responsible for visual perception are at the surface of the retina, such that these nerves are accessible to electrodes. Microchips composed of ordered arrays of microscopic solar cells, capable of converting light into electrical pulses, have been implanted into the eyes of animals and patients. With a 60-electrode array, patients who had been blind are able to recognize objects and read a large print newspaper. (http://www.artificialretina.energy.gov/)

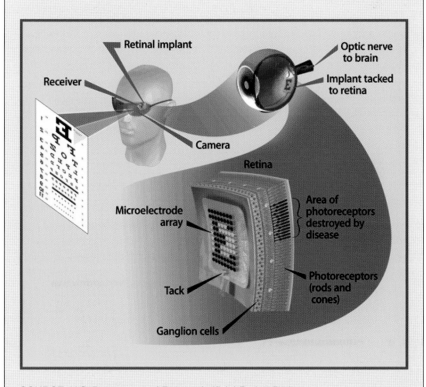

SOURCE: U.S. Department of Energy, Artificial Retina Project.

from a reductionist focus on the basic molecular components of cells, which has uncovered many of the molecular and cellular processes that govern life. Work focusing on other levels of organization—organisms, communities, and ecosystems—also has produced profound new insights.

Reductionism, which dissects and analyzes individual components of living systems to infer mechanisms and to account for the behavior of the whole, has been remarkably successful. And nothing in this report should be interpreted to suggest that support for what is often called "small science" should diminish—indeed it must grow—as the traditional approach to life sciences research will continue to be a major source of discovery and innovation. It has already revolutionized concepts of molecular interactions and cellular functioning, unraveled many of the processes that allow the development of a multi-cellular organism from a fertilized egg, and identified many of the factors that contribute to ecosystem stability. The effort to construct the "parts list" for living systems has been a tremendously exciting intellectual adventure in its own right and has had revolutionary outcomes, such as the biotechnology revolution in medicine and agriculture. Continuing support for peer-reviewed, investigator-initiated research across the broad spectrum of biological sciences is critically important. The New Biology cannot replace—indeed, will not flourish without—those efforts. Construction of an interstate highway does not mean that local road maintenance can cease; the two systems depend on and benefit from each other. Just so, continued support is needed for the science that lays the groundwork for synthesis.

PAST INVESTMENTS ARE PAYING BIG DIVIDENDS

The release in 2000 of the draft sequence of the human genome was the product of a decade-long program that involved billions of dollars in investment. Funding for this project came from multiple U.S. government agencies (especially the National Institutes of Health and the Department of Energy), as well as from international governmental and private sources. In addition, the project was spurred on by competition and collaboration with the private sector, and by the development of new technologies.

The sequencing of the human genome was a goal akin to that of sending humans to the moon, in that the science and technology needed to achieve the mission did not exist when the goal was announced. But new technologies and concepts were developed that have now become routine components of all genome sequencing projects. The magnitude of the challenge spurred creative engagement leading to transformative advances. Many of the advances in sequencing technology were incremental, but there were some game-changing developments, like the demonstration that random shotgun sequencing could be applied successfully to a large complex genome. That kind of transformative event cannot be predicted, but setting an important goal and providing resources to reach it makes it more likely that creative minds will turn to developing revolutionary new approaches in addition to incremental progress.

Random shotgun sequencing, in which a computer detects overlaps of raw sequencing "reads" to construct a complete genome, was not considered useful for human genome sequencing due to the size and complexity of the human genome. However, with the development of new sequencing and computational methods (developed by engineers and computational scientists who turned their efforts to solving biological problems), shotgun genome sequencing became the standard method for genome sequencing and has led to an exponential increase in the number of complete or nearly complete genomes available. This in turn has led to the development of next-generation sequencing technologies that produce massive amounts of sequence data that can only be analyzed computationally.

With the rapid development of sequencing and analysis capabilities, DNA sequencing has become a routine tool in unanticipated areas. A good example is metagenomics, which involves the random sequencing of DNA isolated from environmental samples (such as from soil or water). Metagenomics provides insight into what has previously been a mostly hidden world of microbial diversity, which itself is important because microbes have a fundamental impact on the biogeochemical cycles of the planet and on the health of all its inhabitants (Box 3.6). Another example is population genomics, where researchers are generating multiple complete genomes of different individuals within a species. These data, in turn, serve as valuable input for multiple areas of biology, including genetics, plant and animal breeding, and disease studies. Other biological areas being transformed by genome sequencing include ecology, agriculture, bioenergy research, forensics, and biodefense. None of this would have been possible without the tools and resources developed as part of efforts geared primarily toward sequencing the human genome.

The past 15 years have seen the development of tools and technologies that have extended research capabilities well beyond genome sequencing. These tools include methods to characterize the presence and quantities of many other biological molecules, including transcribed RNA, proteins, metabolites, molecules secreted by cells, DNA methylation patterns, and so on. The comprehensive sets of data generated about these biological molecules—commonly referred to as "omes" (transcriptomes, proteomes, metabolomes, etc.)—are as yet more difficult and expensive to generate and less standardized than genomes. Advances in these technologies will be critical to rapid advances in the life sciences.

Being able to collect and analyze these comprehensive data sets allows researchers to relate and integrate the components of biological systems, a pursuit known as systems biology. They also allow researchers to investigate organisms other than the model systems that have been studied in the past. With relatively little effort and cost, researchers can derive information on an

BOX 3.6
Microbial Genomics

Microbiology, through microbial genomics, is experiencing a renaissance enabled by technological advances over the past several years that have allowed researchers to explore the diversity and metabolic capabilities of a microbial world thousands of times more diverse than before appreciated. This newfound potential is allowing us to understand the critical position that microbes have in the biological world. Harnessing the molecular biology and biochemistry of microbes, either in pure culture under laboratory conditions or in naturally occurring complex communities, promises to contribute significantly to addressing all four challenges presented in this report (Maloy & Schaechter, 2006; Woese & Goldenfeld, 2009). Microbial communities support the growth of plants, affect human health, are critical components of all ecosystems, and can be engineered to produce fuels.

Until the advent of low cost, high throughout sequencing, most of the microbial world was essentially invisible. By necessity, microbiologists focused on the study of individual microbial species grown in pure laboratory culture. Increasingly, it is clear that pure culture does not reflect how microbes live outside of the laboratory and that the microbial world is more diverse, more important, and far more interdependent than had previously been imagined. Interdependence—whereby complex communities of microbes work together to carry out such functions as digesting food, breaking down waste and capturing solar or geothermal energy—is the rule, and the many microbes that can only grow in community were never isolated by classical culturing methods. There is now a tremendous opportunity, and imperative, to develop methods to efficiently characterize these communities. For any given circumstance (e.g., the body of an organism, the soil supporting a specific crop, or the water sustaining temperate fisheries), we must be able to determine the composition of such communities, how they function under conditions that promote the health of the system, and the effects of imbalances in these communities when they are perturbed. The patterns that emerge from these studies can be used to develop predictive models, so that we might recognize problems early and intervene before the situation is irreversible. Integrating microbiology into healthcare, agriculture, energy production, and ecosystem management will be critical to the future of all of these areas.

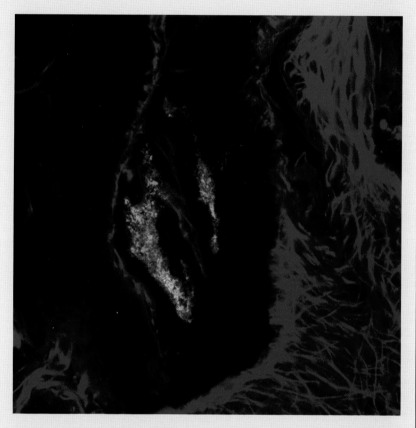

Confocal micrograph depicting the colonization of host animal tissues (blue) by two different types of bacteria (red and green). The bacteria are colonizing extracellularly along the apical surfaces of the host-animal epithelia, in a similar manner to the way that bacteria colonize the mammalian intestine. Unlike the mammalian intestine, which harbors a consortium of hundreds of bacterial types, the animal whose tissue is depicted here, the Hawaiian squid *Euprymna scolopes*, only harbors one species of bacteria, *Vibrio fischeri*. The organ is co-colonized by two strains of *V. fischeri*, a wild-type strain (red) and a mutant strain (green).
SOURCE: Image courtesy of Dr. Joshua V. Troll, University of Wisconsin, Madison.

organism's genome, gene expression patterns, and population variation. One result has been the development of fields such as "polar genomics," "agricultural genomics," and "ecological genomics."

The explosion of unanticipated benefits of the Human Genome Project demonstrates how biology can benefit from large-scale interdisciplinary efforts. Another lesson from the Human Genome Project is that even scientific efforts that appear incremental can spawn transformative advances. Similar efforts to allow systematic characterizations at other levels of biological complexity, like the cell, organism, and community, could have similarly dramatic downstream payoffs.

NEW TOOLS AND EMERGING NEW SCIENCES ARE EXPANDING WHAT IS POSSIBLE

Recent technological advances in a number of fields outside biology make possible unprecedented quantitative analyses of biological systems. These fields are diverse, including physics, electronics, chemistry, nanotechnology, computer science, and information technology. In most instances, tools and methods developed for specific applications in their respective fields have been adapted for use in probing biological systems. But in many cases the complexity of biological systems presents new challenges that call for creative solutions and additional innovation. The descriptions that follow are not meant to be exhaustive or prescriptive. They are examples of the kinds of technologies and sciences that would have major impacts in many areas of biological inquiry.

Foundational Technologies

Information Technologies

Advances in information technology (IT), particularly during the last two decades, have dramatically affected our private lives and all aspects of society. The steady increase in computer power, accompanied by a sharp decrease in cost, has been particularly remarkable. Calculations that would take weeks only 10 years ago and required access to large mainframe computers can now be executed in minutes on a laptop.

The advent of optical fibers with bandwidth of up to 40 Gb/sec (100 Gb/sec predicted for 2010) has enabled increasingly large volumes of data to be seamlessly transferred over the network. The interactive and dynamic manipulation and visualization of complex images has become commonplace, with many ubiquitous applications, most notably in computer games.

It must be realized that implementation of these advances required not only the availability of specialized hardware but very importantly, the development of sophisticated software (e.g. the operating systems and their user friendly interfaces, the algorithms capable of carrying out the computations efficiently, the software protocol for ensuring error-free data transmission, and the standards for data exchange and communication between computers and other devices).

The impact of these developments has been particularly far-reaching for the life sciences, because it has come at the very moment in time when the life sciences are undergoing a historical transition, from a low-throughput descriptive experimental discipline to a high-throughput increasingly quantitative science.

More than ever before, the life sciences are about collecting, archiving, and analyzing information on living organisms and their myriad components, and this effort is distributed across the globe. Worldwide genome sequencing efforts, including the recent efforts to sequence the genome of 1000 individuals, are generating terabytes of sequence data (1TB [terabyte] = 1 trillion bytes) that need to be processed, stored, and analyzed. While information on genome sequences is relatively straightforward to represent because of its one-dimensional nature, it is much more difficult to represent information on the biological function of genes and proteins and their organization into dynamic cellular processes.

Nevertheless, much of the output of life sciences researchers is now captured in electronic form, and databases now include far more than just DNA sequence data. Biological imaging and scanning are producing vast amounts of data about biomolecules, cells, organs, organisms, and environments that is difficult to index and interpret because of its three-dimensional pictorial or even four-dimensional dynamic nature. It is often necessary to preserve such data in its raw form because of uncertainty about how it will eventually be summarized and codified for downstream analyses by diverse users. The sheer size of some of these files suggests that some decisions will be required about what must be saved or made easily accessible. Applications in commerce, Internet search, and data acquisition in the sciences have spurred advances in database systems to handle large data volumes and provide versatile tools for facilitating user interaction, data management, and visualization. Nonetheless, the volume, complexity and diversity of biological data, and the lack of proper conceptual frameworks for representing and analyzing it, will continue to push the limits of data modeling methods and database technology.

In vivo and Real Time Imaging of Cells, Organisms, and Ecosystems

The technologies of in vivo and real-time imaging include a related set of such methods as fluorescence, total internal reflection fluorescence, near-field and confocal microscopy, and functional magnetic resonance imaging, and their related technologies, such as the manipulation of fluorescent proteins, fluorescent dyes, and MRI contrast reagents at the cellular and organismal level.

Cells are densely packed with thousands of interacting components that must be produced, transported, assembled into complexes, and recycled all at the appropriate time and place. Whereas proteomics techniques such as those discussed below aim to provide large-scale systematic characterization of the components of cells or other biological samples, current technology does not allow observation of the spatial and temporal organization of these entities while they are at work in cells. The main limitations to reaching a systems level understanding of living cells is the lack of experimental tools that can analyze the cell's complicated internal complexes as they are forming, working, and disassembling. Several of the experimental tools described below are starting to fill this void; significant progress in this area would be valuable across the life sciences.

Recent advances in imaging techniques, such as cryogenic electron tomography (Cryo-ET), offer the capability of charting cellular landscapes at previously unattainable resolutions of less than 10Å, with predictions to attain near atomic resolution in specific cases (Leis et al., 2009). To date Cryo-ET analyses have been mostly restricted to isolated macromolecular assemblies, small bacterial cells, or thin regions of more complex cells, due to the limited penetration depth of electrons. However, recently developed cryo-sectioning techniques make it possible to transcend these limitations and acquire detailed views of many kinds of cells and tissues. The interpretation of these views relies on help from various techniques for labeling protein constituents (immunolabelling, or use of fluorescent tags), and is a fast evolving area. These techniques can be complemented by information from the rapidly expanding repertoire of known 3D structures of individual proteins, as atomic models of these proteins can be used to expand the lower resolution images obtained by the Cryo-ET technique. This combination of techniques provides unprecedented insight into the molecular organization of cellular landscapes.

Similarly, the technology to characterize the location and activity of individual cells within a living organism is also improving. Substantial progress has been made over the last two decades in extending the application of fluorescent semiconductor nanocrystals (also known as quantum dots or qdots) from electronic materials science to biological systems (Gao et al., 2004). Examples of their recent use in the analysis of biological systems include monitoring the diffusion of individual receptor proteins (e.g., glycine receptors) in living neurons and the identification of lymph nodes in live animals by near-infrared emission during surgery. Multifunctional nanoparticle probes based on semiconductor quantum dots (qdots) have recently been developed for cancer targeting and imaging in living animals. The applications include in vivo targeting studies of human prostate cancer growing in mice and sensitive and multicolor fluorescence imaging of cancer cells under in vivo conditions (Box 3.7). In addition, microfluidic and microfabrication approaches are generating the ability to monitor cells and their components at unprecedented levels of resolution.

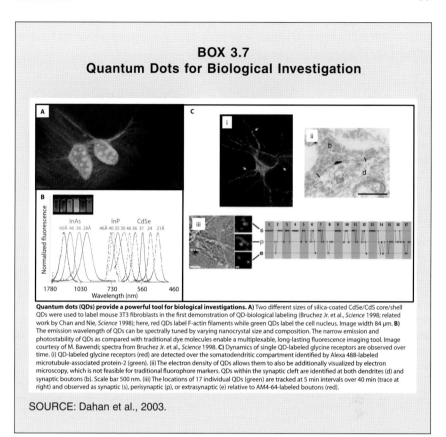

BOX 3.7
Quantum Dots for Biological Investigation

Quantum dots (QDs) provide a powerful tool for biological investigations. A) Two different sizes of silica-coated CdSe/CdS core/shell QDs were used to label mouse 3T3 fibroblasts in the first demonstration of QD-biological labeling (Bruchez Jr. et al., *Science* 1998; related work by Chan and Nie, *Science* 1998; here, red QDs label F-actin filaments while green QDs label the cell nucleus. Image width 84 µm. **B)** The emission wavelength of QDs can be spectrally tuned by varying nanocrystal size and composition. The narrow emission and photostability of QDs as compared with traditional dye molecules enable a multiplexable, long-lasting fluorescence imaging tool. Image courtesy of M. Bawendi; spectra from Bruchez Jr. et al., *Science* 1998. **C)** Dynamics of single QD-labeled glycine receptors are observed over time. (i) QD-labeled glycine receptors (red) are detected over the somatodendritic compartment identified by Alexa 488-labeled microtubule-associated protein-2 (green). (ii) The electron density of QDs allows them to also be additionally visualized by electron microscopy, which is not feasible for traditional fluorophore markers. QDs within the synaptic cleft are identified at both dendrites (d) and synaptic boutons (b). Scale bar 500 nm. (iii) The locations of 17 individual QDs (green) are tracked at 5 min intervals over 40 min (trace at right) and observed as synaptic (s), perisynaptic (p), or extrasynaptic (e) relative to AM4-64-labeled boutons (red).

SOURCE: Dahan et al., 2003.

Also, whole-organism imaging and remote sensing, including satellite remote sensing and multispectral imaging at the ecosystem level, are available in real time. The problems associated with the full development of these methods include image processing and analysis, to enable features to be visualized and automatically recognized. This area is of importance to many of the central problem areas that have been identified, including crop productivity (in analysis of plant cells and growth of tissues) and sustainable crop production, ecological monitoring by ecosystem visualization, and better understanding of human health through advances in medical imaging. To the degree that plant growth is central to biofuel production, these technology platforms are also of importance in this area.

Satellite remote sensing of the earth's surface, beginning in the early 1970s, has dramatically influenced understanding of the distribution of life processes on the planet, as well as pointing critical attention to the rapid human-induced changes at global scales. Broad-spectral reflectance optical sensors, such as the series of Landsat satellites, have been used to measure rates of deforestation, and these and similar sensors are now used routinely to measure these rates over tropical South America. NASA has developed global data sets for the 1980s, 1990s, 2000, and 2005 that cover essentially the entire terrestrial surface in six visible and near-infrared spectral bands at approximately 30 m spatial resolution. This is a powerful time series of the actual change in land-cover and vegetation for the earth's surface, and has already proven useful not only for understanding amounts and rates of deforestation and habitat change, but also for assessing agricultural extent and productivity.

These medium resolution data have been substantially augmented by higher temporal resolution sampling of a broader range of spectral reflectances from sensors such as MODIS on NASA's TERRA and AQUA platforms. MODIS provides twice daily sampling of a wider range of more precise spectral bands, and enables the analysis of long (now nearly a decade) time series of net primary productivity, vegetation distribution, seasonality, surface temperatures, and along with other optical sensors, ocean biological productivity through the measurement of ocean color (i.e., observations of the chlorophyll concentrations of the surface ocean).

Through experimental missions, aircraft missions, and some space observations, remote measurements can increasingly be used to derive specific process-based information, or to retrieve critical parameters directly. Examples include the use of hyperspectral information to detect canopy nitrogen and lignin content, and therefore estimate photosynthetic potential and distinguish individual species distributions, the use of synthetic aperture radar to estimate the distribution of above-ground biomass, and the use of lidars to measure the height distribution of vegetation canopies and thus estimate the vertical distribution of woody biomass, in addition to its total mass.

Earth Science and Applications from Space (National Research Council, 2007b) has identified 17 missions, many of which have specific biological goals related to understanding the interaction of ecosystems, the physical climate system, and human disturbances. These represent the scientific community's best summary to date of the fundamental advances that are believed possible, and that would transform understanding of how ecosystems function now, and how they are expected to function in the future.

High-Throughput Technologies

Recent advances in DNA sequencing technologies have been tremendous. Using current next-generation technology, the Joint Genome Institute, headed by the Lawrence Berkeley National Laboratory and Lawrence Livermore National Laboratory, sequenced over 20 billion nucleotides in the month of October 2008 (DOE Joint Genome Institute, 2009). The ability to sequence individual genomes, or relevant portions of genomes, will have a major impact on the ability to develop and deliver personalized medicines, to speed plant breeding, and to monitor environmental conditions. Already, sequencing costs are so low that they do not represent a barrier to experiments that would have been unthinkable even five years ago. Box 3.8 describes one new sequencing approach made possible by advances in nanotechnology.

Proteins play key roles in virtually all cellular processes. Measuring their expression levels and the chemical modifications that they undergo as a result of changing cellular environments and developmental and disease states has become one of the major goals of present-day biology and medicine. Also, proteins rarely act alone. They interact with one another, often forming large edifices that act as complex molecular machines. The systematic characterization of these interactions is required in order to elucidate the functional interdependencies among proteins.

Technological advances over the past 10 years have made it possible to carry out these various analyses on very large scales, giving rise to the field of proteomics, or the study of all of the proteins in a particular biological sample (for example, a single cell or a drop of saliva). Progress in molecular biology techniques and purification methods, coupled with mass spectrometry (MS) techniques, have played a major role in these advances, with MS increasingly becoming the method of choice for the analysis of complex protein samples.

BOX 3.8
Nanotechnology and Sequencing

There are many competing technologies being developed for DNA sequencing (Shendure & Ji, 2008). One of them provides an illustrative example of the interface between biology and nanotechnology, referred to as single-molecule, real-time DNA sequencing. This method utilizes DNA polymerase, an enzyme that synthesizes DNA, and fluorescent nucleotides (different labels for each of the four nucleotides). Because DNA polymerase incorporates complementary nucleotides, monitoring the fluorescent signal of the added nucleotide during synthesis allows one to determine the DNA sequence of the original DNA. A critical component of the method is the use of a zero-mode waveguide (ZMW), a nanofabricated hole that only allows light to penetrate a tiny distance, so that the fluorescence of a single molecule can be detected despite the presence of high concentrations of fluorescent molecules in the remainder of the sample. A single molecule of DNA polymerase is immobilized at the bottom of a ZMW, which is illuminated from below with a laser light. As each incoming fluorescently labeled nucleotide binds to the DNA polymerase, the signal is detected using single-molecule spectroscopy. The faster, cheaper sequencing that may result from this approach (again, only one of many being pursued) emphasizes the potential impact of collaborations that cross traditional disciplines (here, molecular biology, chemistry, applied physics, and nanoengineering) in the life sciences (Eid et al., 2009).

Of the different MS-based techniques for protein profiling, the most sensitive ones are currently able to detect protein expressed at levels of only a few
hundred copies per cell. MS-based methods for detecting protein interaction
partners and protein complexes have been successful in identifying thousands
of protein interactions and hundreds of multi-protein complexes in simple
organisms such as yeast and bacteria, and are now being extended to higher
organisms such as the mouse and human.

Silicon microelectronics has made computation ever faster, cheaper, more
accessible, and more powerful. Microfluidic chips, feats of minuscule plumbing
where more than a hundred cell cultures or other experiments can reside in a
rubbery silicone integrated circuit the size of a quarter, could bring a similar
revolution of automation to biological and medical research. Using techniques
drawn from engineering, chemistry, and physics, highly miniaturized sensors
and analysis devices can be generated that measure real-time parameters at the
level of individual cells or even subcellular compartments, allowing the study
and manipulation of processes at relevant functional levels.

The expense, inefficiency, and high maintenance and space requirements
of robotic automation systems present barriers to performing experiments. By
contrast, microfluidic chips are inexpensive and require little maintenance or
space. They also need very small amounts of samples and chemical inputs to
make experiments work, making them more efficient and potentially cheaper
to use. These chips are made using optical lithography to etch the circuit pattern into silicon. The etched silicon acts as a mold. Silicone is poured into the
mold and then removed. By stacking several layers of molded silicone and then
encasing them in glass, researchers can create an integrated circuit of channels,
valves and chambers for chemicals and cells—like a rubbery labyrinth.

Cell culture chips with up to 100 chambers have been designed to hold
individual cells and all the microscopic plumbing necessary to add any combination of different chemical inputs to those chambers. Such chips can be
used to test how different inputs might cause stem cells to transform into more
specific cells needed for particular treatments. They could also be used to test
how different combinations of antibiotics affect a particular bacterium. Other
chips can be designed for the preparation of valuable and expensive purified
proteins for structural analysis by X-ray diffraction. The tedious trial-and-error
process of preparing crystals of macromolecules may be greatly accelerated
using microfluidic chips. A recent report describes a new type of microfluidic
chip enabling the detailed analysis of up to a dozen different protein indicators of diseases from a single drop of blood in less than 10 minutes (Chen
et al., 2008). Such chips would significantly lower the cost of clinical lab
tests that measure the presence and relative abundance of specific proteins,
thereby enabling early detection of diseases such as cancer. Microfluidic-based
"sippers" that allow monitoring of cell contents in living systems, devices that
allow sequencing of the genome from a single cell, and multiplexed systems

that monitor parameters in high throughput are all pushing the boundaries of our understanding of the dynamics and complexity of living systems (Box 3.9). These monitoring approaches are also beginning to impact ecological sciences, with real-time 24/7 monitoring of habitat function within reach. In addition, the ability to monitor health parameters from a wristwatch, eyeglass, or even contact lenses is under development for real-time health monitoring and reporting that can occur anywhere.

BOX 3.9
The Chemistrode

The chemistrode is a new microfluidic device created by Rustem Ismagilov and colleagues at the University of Chicago that "sips" from a living cell in a way analogous to that in which a microelectrode measures electrical signals. Using a V-shaped tube with an opening at the point of the V, small amounts of cell contents are delivered to aqueous droplets separated by a hydrophobic carrier, which are then passed through a splitter to create replicate arrays of the contents for downstream analysis. This system has the potential to stimulate, record, and analyze molecular signals in cells (Chen et al., 2008).

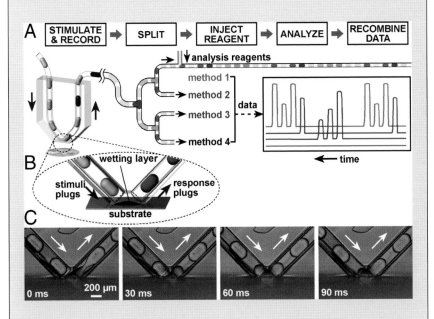

SOURCE: Chen et al., 2008. Copyright 2009 National Academy of Sciences, U.S.A.

Major areas that still require development are nanoscale electrochemical sensors to enable multiplexing with optical sensors, extension of existing sensor technology to broad ranges of analytes, creation of novel platforms for facile deployment, and an increase in the reliability and reproducibility to allow a range of biologically meaningful measurements. The coupling of microfluidics with microfabricated parts will also broaden applicability, and the addition of remote transmission of data will extend the use of these devices outside of the laboratory. In the future, imbedded and largely invisible systems for measurement, analysis, and reporting will become commonplace and will change our lives tomorrow, in much the same way that miniaturized communications technology has changed our world today.

Engineered Biological Systems

Understanding and manipulation of biological systems depends crucially on being able to grow them reproducibly in the laboratory and, for some applications, to scale that ability to commercial production scales. Technological breakthroughs in materials and devices are making it easier to maintain biological entities in an environment that maximizes their production of a particular product or allows their experimental observation and manipulation. Generally these systems are designed to grow cells of a particular type, microbial, plant, or animal, but systems to maintain communities of microbes or support the growth of tissues and organs are the next wave of engineered biological systems.

This capability was developed first for microbial cells in the 1970s and next for animal cells in the 1980s, via in vitro cell culture bioreactors aimed predominantly toward production of therapeutic proteins and non-medical biomolecules such as polymers and specialty chemicals. These reactors included process control systems that could not only track environmental conditions, but also modulate the bioreactors to keep conditions in a desired state. For the most part, however, these bioreactors were limited to cell types that would be productive when growing in fluid suspensions, attached to particle surfaces, or immobilized within membranes.

If tissues and organs from multi-cellular organisms like animals and plants are to be studied in culture, the conditions under which they normally grow must be reproduced—tissues have internal and external surfaces, they often respond to signals from an extracellular matrix, and depend on the continuous delivery of nutrients and removal of waste. Efforts to create effective environments along these lines have accelerated since the early 1990s, in a field generally called "tissue engineering." Tissue engineers develop materials, scaffolds, or devices that provide biochemical and biophysical cues to facilitate cell survival, proliferation, differentiation, and organization into functional three-dimensional tissues. The field of tissue engineering promises to provide more effective experimental systems for studying complex human tissue physiology and pathophysiology in vitro. This capability is highly desirable because animal models fail to capture many crucial facets of human physiology, notably in the areas of tissue-specific transcriptional regulation, drug-induced liver toxicity, pathogenic infection, host immune responses, and cancer. Engineered tissues built with human cells are thus being developed for a range of application areas, including hepatic drug metabolism and toxicity, mammary gland morphogenesis and oncogenesis, lymphoid tissue neogenesis, and stem cell differentiation, and offer promise for scaling to the data collection demands of high-throughput screening and systems biology.

Foundational Sciences

Systems Biology

A crucial requisite for basic science understanding and technology design is a capability for predicting how the entity under consideration will behave under conditions not yet examined. In the New Biology, pursuing and applying this capability to the greatest extent feasible must be a high priority in order to accomplish the kinds of objectives laid out elsewhere in this report. In particular, maximal impact of the molecular and genomic biology revolutions in late-20th century life science will arise from endeavoring to build predictive models of physiological behaviors, in terms of underlying molecular component properties. The lesson from physical and chemical sciences over the past century is that a combination of quantitative multivariate measurement with computational analysis is typically essential for predictive models, and the challenge for life science is that for the foreseeable future it will still have an incomplete knowledge of all of the components and interactions that make up biological systems.

Improved measurement technologies and mathematical and computational tools have led to the emergence of a new approach to biological questions, termed "systems biology," which strives to achieve predictive modeling. Systems biology seeks a deep quantitative understanding of complex biological processes through dynamic interaction of components that may include multiple molecular, cellular, organismal, population, community, and ecosystem functions. It builds on foundational large-scale cataloguing efforts (e.g., genomics, proteomics, metabolomics, etc.) that specify the "parts list" needed for constructing models. The models relate the properties of parts to the dynamic operation of the systems they participate in. The systems approach was applied early on to ecosystem processes (Hagen, 1992), a legacy that has resulted in the development of complex simulation models capable of evaluating interactions among plant communities, ecosystem processes, and atmospheric dynamics.

More recently, systems biology has expanded to molecular components involved in intrinsic cellular processes including gene expression, metabolism, structure and force generation, and regulatory signal transduction. These new advances in systems biology at the cellular level now make it feasible to analyze large data sets of molecular level data that then may be related to phenotypic functions at cellular and higher levels via appropriate kinds of computational methods. A broad range of computational modeling approaches for studying cell signaling and its physiological consequences is needed in the arsenal of systems biology. Fortunately, a wide spectrum of algorithmic methods relevant to systems biology modeling is available from mathematical and computational science, as well as the physical sciences. These tools include Bayesian networks, Boolean and fuzzy logic, inverse modeling, and data assimilation, among others. It clearly can be anticipated that development and application of novel mathematical and computational approaches will be motivated by the difficult problems continuing to arise in systems biology due to issues such as incomplete information concerning system components and properties, heterogeneity and stochasticity, convolution of biochemical and biophysical processes, and the multiple length- and time-scales inherent in attempting to establish predictive models at all levels of biological organization, from the molecular, through the organism, population, ecosystem, and finally, the global scales.

Computational Biology

Biology and mathematics have long been intertwined. The dynamic interplay of hosts and parasites, molecular forces in proteins, biological pattern formation, and signal transmission along axons has been studied using tools of mathematical analysis such as non-linear dynamics and partial differential equations. Fluid dynamics and differential geometry have been applied to heart physiology, group theory to x-ray crystallography, and topological knot theory to the coiling of DNA. From its very origin it was recognized that the study of genetic processes requires probability and statistics. In all these instances the data requirements of the mathematical models were relatively modest, and the daily work of most experimental biologists was relatively unaffected by the results of these studies.

The picture changed completely with the advent of genome sequencing, functional genomics, and systems biology. Biology became an information-based field in which large shared databases are an indispensable tool. The mathematical underpinnings of the field expanded to embrace probabilistic and combinatorial methods. Combinatorial algorithms are essential for solving the puzzles of genome assembly, sequence alignment, and phylogeny construction based on molecular data. Probabilistic models such as Hidden Markov models and Bayesian networks are now applied to gene finding and comparative genomics. Algorithms from statistics and machine learning are applied to genome-wide association studies and to problems of classification, clustering and feature selection arising in the analysis of large-scale gene expression data. The rate of innovation in these statistical disciplines is rapid as new problems of increasing complexity arise in the analysis of models based on heterogeneous data sources. Close collaboration between biologists and mathematicians is increasingly fruitful for both fields by providing new approaches to biological questions and also driving innovation in mathematics.

Synthetic Biology

Another foundational science that reflects the growing role of engineering in biology is synthetic biology. The ability not only to understand, but also to modify and construct biological systems will be essential if we are to apply the power of biology to diverse environmental, energy, and health problems. Synthetic biology aims to use biological modules as the components with which to engineer new biological systems. By standardizing biological parts and the way in which classes of parts can be functionally linked together, this field aims to make large-scale genetic engineering easier and more predictable, potentially leading to cells, organisms, or biologically inspired systems with highly optimized industrial or therapeutic applications.

Synthetic biology is also proving to be an effective teacher as a way to learn more about the fundamental logic of biological systems. Traditionally, natural biological systems have been studied by observation and by dissection (reverse engineering). These approaches alone, however, are often insufficient to uncover the core design principles of a system It can be difficult to identify which components and parameters are most important, especially when dealing with natural systems that have arisen through idiosyncratic evolutionary paths. The ability to build and modify a biological system provides tools to directly probe and interrogate the system. One can modify individual parameters in a controlled and combinatorial fashion to understand which ones are functionally most important and under what circumstances. One can identify minimal or alternative systems that can achieve a particular function, thereby more clearly outlining core design principles. Success in forward engineering is the ultimate test of predictable understanding; failure can be our most constructive teacher. These approaches are already bearing fruit and may ultimately generate the next great conceptual advance: a general understanding of how nature constructs robust and precise systems from noisy and imperfect parts (as well as why these systems fail under certain circumstances).

CONCLUSION

All of these factors—increasing integration within the life sciences and between the life sciences and other disciplines, a deep pool of detailed knowledge of biological components and processes, previous investment in the generation of shared data resources, stunning technological innovations, and crosscutting sciences that are foundational across many applications—have put the life sciences unmistakably on course to a major acceleration of discovery and innovation. It is a matter of great and justified excitement that a sharp upturn in the curve of conceptual progress is coming into view.

But realizing this potential will require a crucial transition within the life sciences. It will require significant investment and will no doubt cause some disruption of engrained educational, institutional, and even intellectual habits. The question must be asked whether the life sciences are ready to capitalize on this potential. Perhaps it would be preferable to continue to focus on current approaches until further progress makes success more likely. What is the urgency, or the claimed opportunity, to move forward now?

One response appeals to America's competitive spirit. The United States was a leader in the development of the life sciences throughout the 20th century and would benefit greatly by remaining in that position in the 21st century. Especially in economically challenging times, the drive to stay at the forefront of critical areas of research can motivate needed investments and changes.

The time to move forward is now.

4

Putting the New Biology to Work

The New Biology approach has the potential to meet critical societal goals in food, the environment, energy, and health, but taking a "business-as-usual" approach to supporting the emerging field will delay achieving its full potential. Success depends on new kinds of investments to enable and drive new, broadly integrated approaches.

SETTING BIG GOALS:
LETTING THE PROBLEMS DRIVE THE SCIENCE

Responses to great challenges often must be enunciated, formulated, and launched before the capabilities to meet those challenges are in place. In this way, the response often motivates the creation of the necessary capabilities. The decisions to send humans to the moon and to sequence the human genome were both made when the relevant technologies were far from being up to the job. In each case, establishing a bold and specific target created unforeseen routes to solutions.

Recent technological and scientific advances have brought the life sciences to a point where rapid progress toward understanding complex biological systems is possible. Many of the essential ingredients are already in place. The New Biology is already emerging, but the interdisciplinary, system-level, computationally intensive projects it encompasses fit uneasily within traditional funding opportunities and institutional structures. A piecemeal strategy, with many different agencies funding interdisciplinary projects and investing in various technologies would continue to advance the efforts of some pioneer researchers whose work has enormous promise. But the cross-cutting technologies and tools that would genuinely empower the New Biology will require significant investment and advance planning. Currently, no mechanism exists for the extremely diverse community of current and future New Biologists to identify, prioritize, and advocate for the investments that would have the biggest impact on the most sectors.

An alternative approach is to set an ambitious goal and invest in the research and technology development needed to meet it. This approach has led to some of America's most spectacular scientific achievements. The committee believes that the best way to capitalize on the unique opportunity presented by emerging capabilities in the life sciences is to undertake a bold national program to apply the New Biology to the solution of major societal problems.

The call for a large commitment to applying the New Biology to big goals is not meant to imply that such a program would consist only of "big science" collaborative projects. The enunciation of big goals is important because it invites the participation of both collaborative groups and individuals from a broad spectrum of disciplines. Solutions to large-scale problems demand contributions from investigators operating both individually and together. Given the need to stimulate both conceptual and technological advances to fulfill the promise of the New Biology, a mixture of both individual and large-scale projects will be necessary. The Institute of Medicine and National Research Council addressed this question in the 2003 report *Large-Scale Biomedical Science* (National Research Council, 2003c). That report states that "the objective of a large-scale project should be to produce a public good—an end project that is valuable for society and is useful to many or all investigators in the field." The report goes on to point out that "large-scale collaborative projects may also complement smaller projects by achieving an important, complex goal that could not be accomplished through the traditional model of single-investigator, small-scale research." The report lists several criteria that characterize projects that are best carried out on a large scale, including external coordination and management, a required budget larger than can be met under traditional funding mechanisms, a time frame longer than that of smaller projects, and strategic planning with intermediate goals and endpoints as well as a phase-out strategy.

The committee chose to focus on four areas of societal need because the benefits of achieving these goals would be large, progress would be assessable, and both the scientific community and the public would find such goals inspirational. Each challenge will require technological and conceptual advances that are not now at hand, across a disciplinary spectrum that is not now encompassed by the field. Achieving these goals will demand, in each case, transformative advances. It can be argued, however, that other challenges could serve the same purpose. Large-scale efforts to understand how the first cell came to be, how the human brain works, or how living organisms affect the cycling of carbon in the ocean could also drive the development of the New Biology and of the technologies and sciences necessary to advance the entire field. In the committee's view, one of the most exciting aspects of the New Biology Initiative is that success in achieving the four goals chosen here as examples will propel advances in fundamental understanding throughout the life sciences. Because biological systems have so many fundamental similarities, the same technologies and sciences developed to address these four challenges will expand the capabilities of all biologists.

The committee suggests that a New Biology approach to the areas of food, the environment, energy, and health will require support for work at different scales, and from basic science to industrial application. As described in chapter 2, the New Biology has the potential to make significant contributions to addressing problems in each of these areas. In each area, the committee has suggested a challenge that is beyond the scope of any one scientific community or federal agency: for food, to generate food plants to adapt and grow sustainably in changing environments; for the environment, to understand and sustain ecosystem function and biodiversity in the face of rapid change; for energy, to expand sustainable alternatives to fossil fuels; and for health, to achieve individualized surveillance and care. The committee's descriptions are meant to be evocative, not prescriptive. The first, and critical, step in designing New Biology programs in these four areas would be to bring to the table all of the stakeholders who could contribute, including scientists and engineers from many different communities, representatives of the relevant federal agencies, and private sector participants from both the commercial and non-profit sector. This step alone—bringing together the diverse talent and resources that already exist and giving them a mandate to plan a long-term, coordinated strategy for solving concrete problems—will already provide significant momentum to the emergence of the New Biology.

The committee does not provide a detailed plan for implementation of such a national initiative, which would depend strongly on where administrative responsibility for the initiative is placed. Should the concept of an initiative be adopted, the next step would be careful development of strategic visions for the programs and a tactical plan with goals. It would be necessary to identify imaginative leaders, carefully map the route from 'grand visions' to specific programs, and develop ambitious, but measurable milestones, ensuring that each step involves activities that result in new knowledge and facilitates the smooth integration of cooperative interdisciplinary research into the traditional research culture.

Implementation of a national New Biology Initiative project does not require creation of a new agency; coordination of the resources already existing in the academic, public, and private sectors is the goal. Estimating the cost of such an Initiative is beyond the scope of this committee, but for the purpose of providing a relative scale, the Interagency Working Group overseeing the National Plant Genome Initiative estimated that the program would require $1.3 billion to fund its programs from 2003 to 2008 ($260 million/year)(NSTC,

2003). The Common Fund, which funds the NIH Roadmap for Biomedical Research, had a budget of $480 million in 2008. Each of these programs has a more limited scope than any of the four proposed New Biology Initiative programs in food, energy, environment and health, so the cost will be too large to be extracted from current research budgets. Whatever the budget, the timeline for such an Initiative must be long enough to justify investing in projects and technologies that will take time to bear fruit—at least ten years. As President Obama said in his address to the annual meeting of the National Academy of Sciences on April 28, 2009:

> As Vannevar Bush, who served as scientific advisor to President Franklin Roosevelt, famously said: "Basic scientific research is scientific capital." An investigation . . . might not pay off for a year, or a decade, or at all. And when it does, the rewards are . . . enjoyed by those who bore its costs, but also by those who did not. That's why the private sector under-invests in basic science—and why the public sector must invest in this kind of research (The White House, 2009).

CROSS-CUTTING TECHNOLOGIES AND FOUNDATIONAL LIFE SCIENCES

A quantum jump in the level at which we understand biological systems will be required to solve these grand challenges. Although there are increasing efforts to apply quantitative approaches to biological questions, more must be done to transform biology from its origins as a descriptive science to a predictive science. We will ultimately be limited in our ability to deploy biological systems to solve large-scale problems unless we significantly deepen our fundamental understanding of the organizational principles of complex biological systems, a staggeringly difficult challenge. The growth of the New Biology will be dramatically accelerated by developing frameworks for systematically analyzing, predicting, and modulating the behavior of complex biological systems. Only with powerful tools to interface with biological systems, accessible to diverse researchers, will it be possible to effectively generate biology-based solutions to the diverse problem areas described in chapter 2.

Many of the foundational technologies and sciences identified as central to New Biology contribute to meeting all four of the critical societal goals. The case for informational technologies is obvious; they will provide the means of disseminating discoveries whether they arise out of research focused on energy, food, environment, or health. Perhaps less obvious is systems biology. Discovering the general principles of dynamic control of the flow of energy, chemicals, and organisms through units spanning from cells to ecosystems is critical for all four societal challenges. The advances in systems biological research will come from insights of computational and physical scientists and engineers as well as cell and molecular biologists. For example, to model the flow of information from the surface of a cell when a hormone stimulates a receptor, to the activation of a set of genes, and, ultimately, cell division requires biologists to establish the experimental system, engineers to measure the time course of changes in thousands of molecules, computational scientists to analyze the data, and all three to integrate the results into a cohesive, testable model. The tools and concepts for each of these steps also have to be created.

The technologies and sciences are highly interconnected. Progress in any of them will support and advance all the others, leading to faster progress in meeting all four goals. Take, for example, the role of synthetic biology in improving pharmaceuticals. Synthetic biologists have already transferred into bacteria all of the necessary molecular machinery to synthesize artemisinin (Martin et al., 2003). This potent anti-malaria compound is naturally produced in small amounts in the leaves of the wormwood tree. Through synthetic biology, the compound can be produced in greater quantity and at lower cost. Clearly synthetic biology has great promise in the area of improving therapeutics and thereby human health. But synthetic biology also has the potential to engineer bacteria that produce high-energy biofuels, thus contributing to the energy challenge; bacterial communities that digest pollutants, thus cleaning the environment; or even sentinel plants that signal the presence of invasive species or crop pests. The field of synthetic biology, however, does not exist in a vacuum; to reach its greatest potential it will require imaging technology to watch individual proteins at work in cells, high throughput technology to measure the output of individual bacteria, engineered biological systems to support high yields of desired products, and information technologies to analyze and model complex metabolic networks.

The foundational sciences and technologies described here are by no means a complete list. In fact, the emergence of new technologies and fields of science as a result of the interdisciplinary collaborations in New Biology is another likely benefit of a major interagency initiative. What is clear is that there are certain technologies and sciences that are of cross-cutting importance and will support communities of researchers whether they are working on food, the environment, energy, or health. Therefore, an interagency initiative will benefit from a mechanism for planning investments in these and other cross-cutting areas. These investments will drive progress in all four problem areas. But advances in these foundational sciences and technologies will not only advance the work of the communities working directly on the New Biology Initiative. The lesson of the Human Genome Project is that these advances will spread into the wider scientific community, multiplying the value and increasing the productivity of researchers throughout the life sciences community. Investment in cross-cutting technologies will make it likely that the United States will be the leader in the resulting new industries with all the attendant economic and job creation benefits.

NECESSITY FOR INTERAGENCY COLLABORATION

Biology-based solutions to major societal problems will not come exclusively from any one area of research. Many federal agencies already support researchers who are pioneers in the development of the New Biology and invest in the cross-cutting technologies and sciences discussed above. But current institutional and disciplinary fragmentation has two consequences. First, traditionally separate research communities often are not aware of the significance of—and therefore do not quickly capitalize upon—advances made in other communities, and second, the multiplying value of investments in cross-cutting technologies and foundational sciences that would benefit all the different kinds of biological research is not readily recognized. Fragmentation within and across institutional structures poses a significant barrier to realizing the full potential of the New Biology. Interagency collaboration will be critical for accelerating the emergence of the New Biology. Through collaboration, the unique strengths of each agency—for example, in technology development, shared facility management, basic and applied research support, or grant review and administration—can be combined to the benefit of all and needless redundancy can be minimized. Most importantly, synergies and entirely new approaches will emerge that would otherwise never have been realized.

Interdisciplinary programs either within or across agencies do exist and some can provide valuable insight into what makes such programs succeed. For example, the Ecology of Infectious Diseases (EID) initiative began in 1999 as a joint program of the National Science Foundation (NSF) and the Fogarty International Center (FIC) of the National Institutes of Health (NIH). The jointly administered program solicits competitive research grants for research on relationships between environmental change and the spread of infectious agents. A 2005 review (Burke et al., 2005) concluded that the program "successfully bridged disparate scientific disciplines and institutional cultures to develop new approaches to critical environmental and health challenges. It has also played an important role in building a cadre of interdisciplinary scientists" and that "the first five years of the EID program have been successful and productive. A total of 34 projects have been funded, and all of them have been both interdisciplinary and appropriately targeted." The review went on to note that—

> The EID program mission overlaps with the missions of several of the NIH Institutes and NSF Directorates. As one of the few joint NIH-NSF programs, the EID program is also a valuable example of effective interagency cooperation. It is to the credit of the program officers and the original partner agencies that the need was recognized and the gap was effectively bridged. It is hoped that the lessons learned from the EID program can help encourage and inform future intra-agency and interagency cooperation.

The report points out management issues that arose from the interagency nature of the program, and made several recommendations that will be even more critical for enabling the larger scale interagency cooperation needed to implement the New Biology. For example, the report recommended that proposal application and reporting processes be streamlined into a single process and that data and sample sharing be promoted. Because of the many lessons learned in implementing this inherently interdisciplinary program, the review suggested that the EID program continue to evolve as a model for interagency cooperation and to strive to include other institutes at NIH and other divisions of NSF.

Another successful interagency program is the National Plant Genome Initiative (NPGI), established in 1998. The NPGI is overseen by the Interagency Working Group (IWG) on Plant Genomes, which includes representatives from NSF, NIH, the Department of Agriculture (USDA), Department of Energy (DOE), Office of Science and Technology Policy (OSTP), Office of Management and Budget (OMB), and, since 2003, the Agency for International Development (USAID). The IWG coordinates all plant genome research activities supported by the participating agencies. In 2008 the NRC issued the report *Achievements of the National Plant Genome Initiative and New Horizons in Plant Biology* (National Research Council, 2008), which evaluated the first five years of the NPGI and made recommendations for the next five-year effort. The report concluded that "NPGI has been very successful by all measures applied in this study" and that "plant genome scientists, as a community, have . . . elucidate[d] basic biological principles that are likely to be broadly operative across plant biology and can thus facilitate rapid applications to crop genomics and improvement." The report also noted that "basic research funded by NPGI to date has served as the springboard for several applied, agency-specific, mission-oriented programs" and that "NPGI principal investigators also reported diverse and substantive translational activities . . . rang[ing] from starting their own companies on the basis of research results to patent filings and licensing arrangements with a variety of plant biotechnology entities."

The NIH Roadmap for Medical Research, which began in 2004, shares many characteristics of an interagency program, although it is confined to NIH. Its goal is to support research that crosses individual Institute and Center missions and to—

> address roadblocks to research . . . by overcoming specific hurdles or filling defined knowledge gaps. Roadmap programs span all areas of health and disease research and boundaries of NIH Institutes and Centers (ICs). These are programs that might not otherwise be supported by the NIH ICs because of their scope or because they are inherently risky. Roadmap Programs are expected to have exceptionally high potential to transform the manner in which biomedical research is conducted (NIH, 2009).

The first round of funding included support for Interdisciplinary Research consortia, Clinical and Translational Science awards, projects in nanomedicine and structural biology, and centers for biomedical computing, and networks and pathways, among others. The second round of funding, in 2008, added epigenomics and analysis of the human microbiome. No external review of the program has taken place comparable to the NRC review of the NPGI, but many of the projects funded in the first round of funding were given renewed support for a second term, and internal evaluation of the effectiveness of the programs is built into the program.

The joint effort between DOE and NIH to make synchrotron resources available to the life sciences research community is another example of successful interagency collaboration. The DOE funds the building and operation of the synchrotron facilities and NIH funds the building and operation of beamlines and experimental stations specifically designed for life sciences applications. These collaborations have been especially important for structural biology and support for life sciences research is an increasingly important part of DOE's portfolio. Currently ">40% of all research done at synchrotrons is in the biomedical sciences, although synchrotrons were originally developed for high energy physics experiments" (National Center for Research Resources, 2009).

The effort required for success in meeting the four major societal goals is different in scale from NPGI, EID, or the NIH Roadmap: it will need to involve more agencies, a larger investment and a long-term commitment. True interagency collaboration will demand interagency strategic planning (including a commitment to supporting the development of novel, integrated approaches to education), interagency funding, and interagency evaluation and review. Such an infrastructure for interagency collaboration will empower and enable the joint efforts of individuals and groups who are currently insulated from one another by bureaucratic barriers. Importantly, the need is not for a new agency, which would merely establish another silo, or even for a reorganization of existing agencies, but rather for mechanisms that actively permeate their current boundaries. Successful "permeation" would bring together scientists with different backgrounds, expertise, and goals, sparking new shared visions and synergies that could not have been realized separately, new ways of conceiving and addressing major societal challenges, and eventually, transformational advances (Box 4.1).

It is worth emphasizing explicitly three elements essential for achieving these ends. First, availability of dedicated interagency funds, outside of each agency's individual budget needs, will motivate their involvement. Second, interagency strategic planning will place the focus of a broad spectrum of scientists and engineers on discovering novel, shared, life sciences-based approaches to these societal goals. Inclusion of some private sector scientists in this integrated planning effort might evolve novel public-private partnerships that could help drive late-stage efforts. And finally, given the cross-cutting and interdisciplinary nature of the science that will be needed, establishment of a common interagency peer-review and evaluation process will set shared standards of

---

BOX 4.1
How Might Interagency Programs
Catalyze the New Biology?

Most key scientific advances to date have been funded by disciplinary funding programs. Advances in the New Biology will require programs that compel integration across disciplines, and synthesis that allows fundamental biology to be applied to key social challenges. One example of a funding program that stimulates such integration of computer science, informatics, and biology is the National Science Foundation–supported National Center for Ecological Analysis and Synthesis (NCEAS), located at the University of California-Santa Barbara. NCEAS provides long-term support in ecoinformatics, with on-site expertise in mathematics and geospatial modeling, visualization, and data synthesis, and invites individuals and teams to assemble at the center to conduct new kinds of research (NCEAS, 2009).

The United Kingdom's Engineering and Physical Sciences Research Program provides another approach to stimulating interdisciplinary scientific research through "sandpits." Sandpits are residential workshops that include 20–30 participants from multiple disciplines, who work together to develop new research projects. By providing an opportunity for exploring possible collaborations and immediate feedback on proposals, sandpits aim to "drive lateral thinking and radical approaches to addressing particular research challenges" (EPSRC, 2009). These efforts are resulting in dramatic advances. Similar efforts by interagency programs could launch the New Biology in the United States.

---

excellence and drive periodic assessments of progress. A national New Biology Initiative would provide all three of these elements.

There is every reason to expect that just as the Human Genome Project (HGP) had an impact across the life sciences far beyond the sequence data generated, investments in problem-focused projects and foundational technologies and sciences will have similarly profound effects. The HGP had the advantage of a clear and definable endpoint—the complete sequence of the human genome—and a similar endpoint for some of these interdisciplinary and cross-cutting projects may be more difficult to define. However, the success of the HGP justifies community-wide efforts to plan and implement strategies to address challenges in the areas of food, the environment, energy, and health, and to invest in those technologies whose development would most significantly contribute to the success of those programs.

THE ESSENTIALITY OF INTERDISCIPLINARY COLLABORATION

The New Biology depends on interdisciplinary collaborations among scientists and engineers who share sufficient common language and understanding to envision and embrace common goals. To expand the pool of such individuals, it will be important to educate students in new ways. Interagency funding mechanisms could give universities incentives to create novel interdisciplinary entities that provide the basis both for new research approaches and for new educational strategies.

Research universities and academic medical centers have for hundreds of years been structured around departments and colleges that circumscribe specific disciplines and intellectual approaches (National Academies, 2004). These structures have had enormous value in encouraging discovery, establishing sufficient focus to virtually define whole fields, and imparting increasingly refined expertise to successive generations of trainees. Indeed, it is in many ways due to the success of these delineated departmental structures that the base of knowledge in each field has advanced sufficiently to make each relevant and potentially contributory to the others. Analogous to the separate government agencies, however, traditional department structures also serve as bureaucratic barriers that inhibit communication and productive interaction. Traditional metrics of success are accomplishments that can be ascribed to individual units, including grant generation, buildings, laboratory equipment acquisition, and financial support for faculty. Faculty within these units being considered for tenure and promotion are reviewed within the department structure, leaving them vulnerable if their focus is interdisciplinary. Certain institutions have recognized these limitations of traditional departments for establishing the New Biology, and have responded not by eliminating departmental structures, but rather by supplementing or overlaying them with interdisciplinary programs or institutes that have both research and educational objectives. Examples include the QB3 Institute at the University of California, Berkeley and the Institute for Bioengineering and Bioscience at the Georgia Institute of Technology.[1] The availability of interagency funds targeted to foster and nurture such integrative programs and institutes would strongly incentivize universities to establish and maintain them, and could prompt a reframing of promotion standards that recognizes the value of collaborative and interdisciplinary education and research in the life sciences.

THE CENTRAL IMPORTANCE OF INFORMATIONAL TECHNOLOGIES IN ENABLING THE NEW BIOLOGY

Information is the fundamental currency of the New Biology. Interagency collaboration to develop the information sciences and technologies necessary to handle biological data would make the single largest contribution to future

---

[1] For more information, see the websites of the California Institute for Quantitative Biosciences (http://research.chance.berkeley.edu/page.cfm?id=56) and the Parker H. Petit Institute for Bioengineering and Bioscience at Georgia Tech (http://www.ibb.gatech.edu/).

life sciences research productivity. Provision of resources for the transmission, exchange, storage, security, analysis, and visualization of biological information will be essential. Biological research is increasingly supported by large-scale information resources available over computer networks. The development of these resources is a community effort. Researchers provide data acquired in their own laboratories, and data management systems organize the shared data and provide software tools for accessing, displaying, and interpreting parts of the data.

Traditional dissemination of results through publications in journals can convey only a fraction of the information that is generated in most experiments. To capture the full benefit of funded scientific work, one must maximize the ability to share that information. Information about research results that is not made accessible is lost to the rest of the research community and thus can be considered a hidden tax on scientific research funding. Ongoing support for the storage, curation, and accessibility of data is critical, but it is also exceedingly expensive and, for funding agencies, comes at the expense of funding new research. Many specialized communities already exist to support database resources, for example, those focused on model organisms, such as Fly Base for the *Drosophila* community and TAIR for the *Arabidopsis* community. There will be increasing demand for resources to support these efforts, especially to support coordination among these specialized communities.

Because so much can be learned in biology by comparing results across different organisms and systems, biological data have more value if made available in a form that can be easily shared, meaning that measurements from one laboratory to another need to be clearly defined. Ideally, biological and biomedical experiments should adhere to nomenclatures and protocols specified by standards bodies in consultation with communities of researchers. As much as possible, data should be reproducible, with no ambiguity as to their meaning and the experimental conditions under which they were acquired. However, rigid application of standards can hold back the introduction of new technologies and their application to a widening range of environments and conditions. Innovation in experimental technologies and their application will inevitably outpace efforts at standardization. Thus, while enforcing standards for mature technologies, data management systems must also incorporate diverse types of ad hoc experimental measurements that are informative even if loosely characterized.

A more recent development in the life sciences is the potential to derive new information from existing collections of data. Hypotheses can be tested and connections across different biological systems discovered using data acquired from the published literature by curation or automated search, transferred from other databases, or inferred from experimental data by various forms of aggregation, classification, clustering, comparison, annotation, or even analogical reasoning. For example, most of the reported assignments of proteins to functional categories are derived not by biochemical experiments, but by imputation from the functional classification of similar proteins, often from a different species. The value of existing data can be multiplied by these approaches, but the basic scientific requirement of reproducibility requires that database management systems provide tools enabling researchers to trace the origins of such indirect inferences and assess the supporting evidence.

The study of complex biological problems typically requires the integration of diverse data sources (Box 4.2). For example, understanding the possible

---

BOX 4.2
The Critical Role of an Information Infrastructure:
Two Examples

Electronic Medical Records

The revolution in information technology has provided an enormous opportunity to make electronic medical records (EMR) a reality. These records not only have the potential to improve the quality of health care, but also could contribute substantially to basic biomedical research. In his speech to the annual meeting of the National Academy of Sciences, President Obama noted that EMRs offer "the opportunity to offer billions and billions of anonymous data points to medical researchers who may find in this information evidence that can help us better understand disease" (The White House, 2009).

The great potential of an EMR for biomedical research is that it provides integrated health information, demographic data, imaging, and laboratory results for each individual. Currently, all the information that resides in each individual's medical records is essentially invisible to researchers. The ability to search this massive data source would allow researchers to detect patterns in drug side effects, relationships between genomic information and disease incidence, spread of infectious diseases, and many others. However, the power of this resource to drive discovery and improve health has yet to be realized. This tremendous opportunity depends on developing an adequate information infrastructure.

Turning all of the information in patients' medical records into a form that can be standardized, digitized, secure, and anonymous is a major challenge and will require developing adequate network and analysis capabilities so that researchers can make full use of the data. The range of useful information that could be included is already vast and will only grow with time as the affordable genome, high-throughput proteomics and metabolomics technologies, and ever more sophisticated imaging are just over the horizon. A major effort to standardize (and anonymize) an EMR that provides the full complement of patient information, and to develop the resources to make those anonymized records fully accessible to researchers, would be an enormous boon to clinical research.

impact of a new cancer drug might involve data from human genome-wide association studies, experiments with mice, characterizations of known molecular pathways and metabolic processes in yeast, and clinical experience with related drugs. A database on biodiversity might contain genetic sequence, photographs, movies, museum catalogues and digital representations of samples, geospatial coordinates, satellite images of collection sites over time, and detailed information about range, habitat, or behavior; ideally all of these kinds of data would be cross-referenced. Statistics, machine learning, and data mining, supported by

The National Ecological Observatory Network

It has become increasingly evident that long-term measurements of ecological function are key to sensing critical changes in the environment. Many studies are now revealing that short-term assessments are not capable of revealing such diagnostic and critical trends as changes in lake ice-melt, glacial melting, or changes in seasonal behavior that signal biological responses to climate change. Research platforms that allow early detection of changes in biological functioning over continental scales are necessary not only for understanding the interactions of climatic change and land use with ecological processes, but also for anticipating threshold responses that could occur during rapid environmental change. Similar to monitoring human health over broad scales to assess trends in risk and improvement, regular collection of data on ecosystem carbon dioxide exchange, land use change, and invasive and other species distributions is critical for understanding and predicting future conditions that influence human well-being. The currently planned NEON research platform represents such an initiative.[a] Just as with EMRs, the kinds of data researchers will need to share, compare, and analyze are exceedingly diverse: satellite images; air, water and soil characteristics; measurements of species diversity and population sizes; changes in the genomes, health, and behavior of organisms; and many others. The benefits of such a data resource to understand, monitor, and predict environmental conditions would be great.

Achieving either of these goals—making the information collected in EMRs available to clinical researchers or the environmental information provided by NEON and many other sources to ecologists—will require a concerted effort. Ideally, even those two very different sets of data would benefit from being inter-operable; understanding the impact of the environment on human health, or how infectious agents pass between animals and humans would be just two possible applications. The full benefit of the impending revolutions in the life sciences discussed in this report will require a national effort to develop an information infrastructure that would support these applications.

---

[a] For more information, see the National Science Foundation report, The NEON Strategy (http://www.neoninc.org/sites/default/files/NEON.Strategy.July2009.Release2_2_0.pdf).

advances in probabilistic modeling, computational simulation, discrete mathematics, algorithms, and data structures, are rapidly advancing in their ability to extract more information from such complex data sets. Innovative methods of information display, based on advanced graphics capabilities including animation and virtual reality, will be essential for biological researchers to visualize such complex models.

As argued throughout this report, the fundamental unity of biology means that data generated to develop biofuels are relevant to biomedical researchers and vice versa. Thus, building a system that captures the most possible value from ongoing research is a challenge that must be addressed above the level of any single biological subdiscipline or any one funding agency. The value of providing a standardized, shared database with a user-friendly interface is exemplified by Genbank (National Center for Biotechnology Information, 2009), which provides researchers with a steadily increasing database of sequence information and standardized tools such as BLAST with which to analyze it. Genbank is housed within NIH, but its use is cross-agency. Every biology-related publication is required to deposit any sequence generated into this central sequence database, and biologists funded by every agency make use of it. The Genbank model has not been achieved for other types of data that may not be as easy to share and standardize as sequence information. But this does not make such data less important.

There is no single, obvious solution to the challenge of providing a flexible, efficient, and high-performance information infrastructure for the data that will power the New Biology. As technology and biological knowledge advance, both requirements and capabilities will shift. But explicitly acknowledging the essential role of information to the life sciences and investing the effort and resources necessary to develop robust informational technologies and sciences would have an enormous pay-off in capturing the full value of life sciences research results. An immediate, interdisciplinary, and interagency effort to address the information requirements of the New Biology would provide a system-wide solution to a problem that is imposing greater and greater costs on the life sciences research establishment.

ENGAGING THE PRIVATE SECTOR IN THE NEW BIOLOGY

The private sector has a great deal to contribute to the proposed New Biology Initiative and should be engaged in the review and assessment of these interdisciplinary projects. Both commercial and non-profit entities will be helpful in assessing knowledge in the field, helping to set objectives and evaluating progress. In some areas, the private sector has capabilities more advanced than the public sector, and is setting the standards for the field (e.g., handling of data for use by web-based search engines). In such cases, interdisciplinary projects would benefit from involving the private sector not only in review and assessment, but also as explicit participants in the projects. In other areas, the private sector has data that are not readily available in the public domain that could be included in the projects. For example, efforts in the analysis of biological signaling pathways and networks rely on the compilation of extensive experiments involving manipulation of gene expression. The pharmaceutical industry has a large amount of such data, based on small-molecule drug manipulation, that is not readily available in the public domain. Similarly, there is a large amount of data from genetically informed breeding experiments in the agriculture industry. Such data would greatly facilitate the advancement of these pre-competitive opportunities, which would serve to benefit all stakeholders.

EDUCATING THE NEW BIOLOGIST

To thrive, the New Biology will require researchers with both depth of knowledge in a specific discipline and highly developed computational and quantitative skills. In addition, the New Biology will require these investigators to be well versed enough in varied disciplines and technologies to facilitate dialogue with other researchers and participate in integrated research.

The emergence of the New Biology signals the need for changes in how scientists are educated and trained. A highly visible science program like a New Biology Initiative could inspire a new generation of students to see becoming a scientist or engineer as a way to contribute to solving important societal problems. The Initiative itself would provide the opportunity to put in place and evaluate new educational and training opportunities.

Thousands of reports, surveys, public speeches, articles, and television shows have bemoaned the quality of science education in the United States and numerous solutions to poor performance have been proposed. Many of these solutions would contribute to preparing students for careers as New Biologists. Implementing these solutions will require investment in human resources and materials and interaction among educators and researchers from a broad spectrum of disciplines.

The New Biology represents an integrated, problem-focused approach to science that is entirely consistent with research on how students learn best. Just as the goal of landing on the moon inspired a generation of students, high visibility projects using biology to solve important problems could provide a platform to engage all students in the process of science, and illustrate the excitement and benefits of using science and engineering to solve problems. An ambitious, high visibility program would demonstrate that basic science research is not distinct from society but is a critical ingredient in developing innovative solutions to societal problems.

Integrating information from several disciplines to study practical questions is a valuable approach at any educational level from kindergarten on. But it is at the undergraduate and graduate levels that the New Biology both demands and presents an opportunity for new approaches. The New Biology makes it clear that biology is not only about observing and describing natural history and phenomena. Rather than teaching each level of biological organization separately—from molecules to cells to organs, etc., and on to ecosystems (if time allows)—a New Biology curriculum would emphasize the interconnections among those levels to understand system-level phenomena. Harvard University, for example, has recently introduced introductory courses that teach basic science material in the context of understanding AIDS treatment, or the possibility of synthetic life (Box 4.3).

Such an approach makes it clear that quantitative analysis, physics, and chemistry are necessary to understand complex issues, along with biology. As

---

BOX 4.3
Connecting Bio 101 to Real-World Issues:
An Interdisciplinary Approach

In 2005–2006, Harvard University launched two semester-long introductory courses that provide an interdisciplinary introduction to biology and chemistry. The first course synthesizes essential topics in chemistry, molecular biology, and cell biology and the second synthesizes essential topics in genetics, genomics, probability, and evolutionary biology. Scientific facts and concepts are introduced in the context of exciting and interdisciplinary questions, such as understanding the possibility of synthetic life, the biology and treatment of AIDS and cancer, human population genetics, and malaria. Through interdisciplinary teaching, students' grasp of fundamental concepts is reinforced as they encounter the same principles in multiple situations.

Each course is taught by a small team of faculty from multiple departments. Members of each teaching team attend all lectures and participate for the entire term. The preparation for and teaching effort in each course offering is integrated. Teaching assistants are also drawn from different departments and work in small interdepartmental teams.

Development of these courses required institutional support. The President, Dean of the Faculty, and the Chair of the Life Sciences Council all provided funds to support a one-year curriculum development effort, lab renovations, lower teaching fellow-student ratios, equipment, and development of teaching materials. One of the founding faculty member's HHMI undergraduate education award contributed to developing specific sets of teaching materials.

Success depended on finding faculty members with personal commitments to the principles of the courses and willingness to work as a team to build the new courses from scratch. This effort was rewarded as individual departments agreed to count these interdepartmental and interdisciplinary courses toward their respective departmental teaching expectations.

Since the courses were implemented, undergraduate enrollment in introductory life sciences courses is up more than 30 percent and the number of life sciences majors has risen 18 percent.

students are taught to approach science as an exercise that solves a problem, they will recognize how mathematics, physics, chemistry, computational science, and engineering contribute to the problem-solving process and therefore see the relevance of and be more motivated to master these other disciplines. Students and teachers alike will recognize that memorization of observations and facts do not allow one to understand or predict how complicated biological systems behave—and without that ability one will not be able to solve problems.

Engaging students in the New Biology will require science teachers who understand and can pass on the interdisciplinary nature of science problem-solving. Exciting undergraduate experiences that are science based will not only help attract students into research careers, but also equip those life science majors who choose teaching careers with the disciplinary knowledge and hands-on experience to teach the New Biology.

Many of the changes that would help prepare students to practice the New Biology have been recommended in several previous reports (Box 4.4), especially a 2003 NRC report, *Bio 2010: Transforming Undergraduate Education for Future Research Biologists* (National Research Council, 2003a). Bio 2010 recommended that each institution of higher education reexamine its current curricula and ensure that biology students gain a strong foundation in mathematics, physical

---

BOX 4.4
Previous Reports Evaluating Science Education

1. *Transforming Undergraduate Education in Science, Mathematics, Engineering, and Technology* (National Research Council, 1999)
2. *Evaluating, and Improving Undergraduate Teaching in Science, Technology, Engineering, and Mathematics* (National Research Council, 2003b)
3. *Learning and Understanding: Improving Advanced Study of Mathematics and Science in U.S. High Schools* (National Research Council, 2002)
4. *America's Lab Report: Investigators in High School Science* (National Research Council, 2006)
5. *How People Learn: Mind, Brain, Experience, and School* (National Research Council, 2000b)
6. *How Students Learn: Mathematics in the Classroom* (National Research Council, 2005)
7. *Bio 2010: Transforming Undergraduate Education for Future Research Biologists* (National Research Council, 2003a)
8. *Fulfilling the Promise: Biology Education in the Nation's Schools* (National Research Council, 1990)
9. *Educating the Engineer of 2020: Adapting Engineering Education to the New Century* (National Academy of Engineering, 2005)
10. *Math and Bio 2010: Linking Undergraduate Disciplines* (Steen, 2005)

and chemical sciences, and engineering as biology research becomes increasingly interdisciplinary. Concepts, examples, and techniques from mathematics, and the physical/chemical sciences should be included in biology courses, and biological concepts and examples should be included in other science courses. College and university administrators, as well as funding agencies, should support mathematics and science faculty in the development or adaptation of techniques that improve interdisciplinary education for biologists. Bio 2010 also called for laboratory courses to be as interdisciplinary as possible, and for all students to be encouraged to pursue independent research as early as is practical in their education. Finally, the report recognized that faculty development is a crucial component to improving undergraduate biology education and called for efforts to provide faculty the time necessary to refine their own understanding of how the integrative relationships of biology, mathematics, the physical/chemical sciences, and engineering can be best melded into either existing courses or new courses in the particular areas of science in which they teach.

Implementing the recommendations of the Bio 2010 report would go a long way toward preparing the biology students of the future to practice the New Biology. The advances in life sciences research described in this report will create tremendous opportunities for students in the coming decades. Both at the undergraduate and graduate level, a new generation of students could learn in different ways, be challenged by new curricula and approaches, and contribute to breakthroughs that can barely be imagined today. Implementation, however, requires resources, time, and flexibility on the part of university administrators, faculty, and even students, who must be convinced that interdisciplinary courses will satisfy graduate school or medical school admissions requirements. A national New Biology Initiative could have a lasting impact by devoting some of its resources to providing incentives for universities and researchers to find innovative ways to implement recommendations like those in Bio 2010, and to identifying and disseminating best practices. Grants programs could support development of interdisciplinary courses like Harvard's introductory biology courses, or Princeton's integrated science curriculum[2] at other institutions. The National Academies Summer Institute on Undergraduate Education in Biology, created in direct response to a *Bio 2010* recommendation, is another approach that could be expanded with additional funding (Box 4.5).

A national New Biology Initiative could also support graduate training programs designed to prepare researchers for careers as New Biologists. The Integrative Graduate Education and Research Traineeship (IGERT) program is an example of such a program (Box 4.6).

The new biology will be most successful if it attracts the best students from a wide range of backgrounds. Communicating the excitement of biological

---

[2] More information can be found at http://www.princeton.edu/integratedscience/.

BOX 4.5
National Academies Summer Institute on
Undergraduate Education in Biology

The National Academies Summer Institute seeks to transform undergraduate biology education at research universities nationwide by improving classroom teaching and attracting diverse students to science (Handelsman et al., 2004; Pfund et al., 2009). Teams of two or three faculty members, most of whom teach introductory courses, learn about and implement the themes of "scientific teaching" (Handelsman, et al., 2004)—active learning, assessment, and diversity—during a week-long workshop dedicated to teaching and learning. Participants work together to develop materials and lessons that they agree to implement in their courses in the following year.

The impact of the Summer Institute is far greater than the individual teaching materials; it transforms how individual faculty members view their teaching and, by extension, influence other members of their departments and their disciplines to make similar transformations (Pfund et al., 2009). Participants are asked to disseminate what they learn at the Institute with colleagues on their campuses, and university administrators commit to support participants in tangible ways upon their return to campus. Participants are named National Academies Education Fellows in the Life Sciences and are encouraged to become ambassadors for education reform on their campuses and throughout their professional communities. The aim is, therefore, to leverage a program that directly reaches 40–50 faculty per year—who themselves teach 15,000–25,000 students per year—into one that reaches hundreds of thousands of students.

Since 2004, more than 250 instructors from 82 institutions in 40 states have participated in the Summer Institute including a broad cross-section of faculty from throughout all of biology—anatomy to zoology—as well as deans and department chairs, curriculum and laboratory coordinators, lecturers and postdocs.

The Summer Institute is supported by the Howard Hughes Medical Institute, the Research Corporation for Science Advancement, the Burroughs Wellcome Fund, the Presidents' Committee of the National Research Council, and the University of Wisconsin–Madison.

---

research is crucial to attracting, retaining, and sustaining a greater diversity of students to the field (Box 4.7).

All of the agencies that support life sciences research have implemented programs to attract participants from underrepresented groups. Some groups remain underrepresented—in 2005, African Americans received 3.6 percent and Hispanics 5.2 percent of doctorates in the biological sciences (Hill, 2006)—it is certain that there is much to be learned by studying the effectiveness of these different programs. A recent NRC workshop summary *Understanding Interventions that Encourage Minorities to Pursue Research Careers* (National Research Council, 2007a) discussed the need for research efforts to identify the

---

BOX 4.6
The Integrative Graduate Education and
Research Traineeship Program

Federal agency funding programs can be very effective at stimulating entre-preneurship and change at academic institutions. For example, the National Science Foundation supports large grants through its Integrative Graduate Education and Research Traineeship (IGERT) program. This program has catalyzed numerous universities (currently 125) to advance interdisciplinary education. According to the program website—

the IGERT program was developed to meet the challenges of educating U.S. Ph.D. scientists, engineers, and educators with the interdisciplinary backgrounds, deep knowledge in chosen disciplines, and technical, professional, and personal skills to become in their own careers the leaders and creative agents for change. The program is intended to catalyze a cultural change in graduate education, for students, faculty, and institutions, by establishing innovative new models for graduate education and training in a fertile environment for collaborative research that transcends traditional disciplinary boundaries.

The IGERT program, which has supported almost 5,000 students since its inception in 1998, is an example of how federal agencies can catalyze change within U.S. institutions to reach true interdisciplinarity.

---

elements that characterize programs that are successful in increasing minority participation. Including resources in a New Biology Initiative to encourage minority participation and, importantly, to evaluate the success of those efforts, will have an important pay-off in ensuring that the new biology benefits from all of the talent this diverse country has to offer.

CONCLUSION

Many intellectual, technological, and institutional challenges will need to be met in order for the New Biology to reach its full potential. Perhaps the most challenging step will be achieving widespread recognition that an integrated approach to solving problems with the life sciences is important and worthwhile. Some portion of the life sciences research enterprise will need to be devoted to approaching the science in this new way. Empowering the New Biology means adding a new layer to the traditional approach; an approach that is purposefully organized around problem-solving; marshalling the basic science, teams of researchers, technologies, and foundational sciences required for the task; and coordinating efforts to ensure that gaps are filled, problems

BOX 4.7
The International Genetically Engineered Machine (iGEM) Competition

Every November for the last five years, hundreds of dedicated young synthetic biologists from around the world have gathered at MIT for the annual iGEM competition. Modeled after popular robotic design competitions, iGEM brings together teams of students whose challenge is to use standard biological parts to design and build a novel genetic-encoded machine that carries out an interesting or useful function. In 2008, 84 teams from over 20 countries participated. Most teams are from undergraduate colleges and universities, but more recently, high school teams have also begun to participate.

The iGEM competition has become a major force in both education and innovation. Scores of top students are drawn to the excitement of the new field of synthetic biology, a field that is revolutionizing how biological systems are viewed and has the potential to solve many societal problems. Students come from biology, computer science, engineering, and many other fields, but work together to formulate their own projects. All the standard biological parts they design are submitted to the iGEM parts registry and projects are described on open websites. The teams gather at MIT to present their work to a panel of judges.

iGEM projects rival those of professional research laboratories and biotech companies in sophistication, and frequently exceed them in innovative thinking. Projects have included design of bacteria that sense arsenic, a bacterial replacement for blood, and a synthetic cellular organelle that could be used to house biofuel pathways.

The iGEM competition provides a model for future modes of education in biology. Unlike many summer research projects, iGEM projects are emergent—students come up with their own ideas and work as a team to develop and execute them. The creative challenge, competitive framework, and required self-investment results in extraordinary levels of motivation and innovation. The forward-looking iGEM team projects may foreshadow how biology will be practiced and applied in the future.

solved, and resources brought to bear at the right time to keep the project moving forward. Close interaction between larger, problem-oriented projects and the more decentralized basic research enterprise will be critical—and mutually beneficial—as discoveries will continue to emerge from traditional approaches, and advances that benefit all researchers will emerge from the large projects. The New Biology Initiative would represent a daring addition to the nation's research portfolio, but the potential benefits are considerable: an immensely more productive life sciences research community; new bio-based industries; and, most importantly, innovative means to produce food and biofuels sustainably, monitor and restore ecosystems, and improve human health.

5

Recommendations

Society is at a tipping point in terms of challenges that influence our collective long-term future; the United States is well-poised to exert leadership in addressing these urgent needs by creating a New Biology. Our response will require launching an ambitious new effort that empowers individuals, agencies, academic institutions, and the private sector to integrate a deeper understanding of biology with practical applications across the agriculture, environmental, energy, and health sectors. Reaching these goals will require investing in a new approach to setting research goals—taking a long-term and cross-cutting approach both to foundational science and technology development, and to evaluating progress and outcomes. A key factor in meeting these challenges will be putting in place a flexible and functional interface among all of the agencies whose programs touch on the life sciences. The committee believes that the most promising way to achieve this interface is through a national initiative. It will be individual agencies, however, that implement components of the initiative; therefore developing an efficient management and oversight structure to facilitate communication and shared decision-making on cross-cutting investments is critical.

Finding 1
- The United States and the world face serious societal challenges in the areas of food, environment, energy, and health.
- Innovations in biology can lead to sustainable solutions for all of these challenges. Solutions in all four areas will be driven by advances in fundamental understanding of basic biological processes.
- For each of these challenges, solutions are within reach, based on building the capacity to understand, predict, and influence the responses and capabilities of complex biological systems.

- There is broad support across the scientific community for pursuing interdisciplinary research, but opportunities to do so are constrained by institutional barriers and available resources.
- Approaches that integrate a wide range of scientific disciplines, and draw on the strengths and resources of universities, federal agencies, and the private sector will accelerate progress toward making this potential a reality.
- The best way for the United States to capitalize on this scientific and technological opportunity is to add to its current research portfolio a New Biology effort that will accelerate understanding of complex biological systems, driving rapid progress in meeting societal challenges and advancing fundamental knowledge.

Recommendation 1
The committee recommends a national initiative to accelerate the emergence and growth of the New Biology to achieve solutions to societal challenges in food, energy, environment, and health.

Finding 2
- For its success, the New Biology will require the creative drive and deep knowledge base of individual scientists from across biology and many other disciplines including physical, computational, and earth sciences, mathematics, and engineering.
- The New Biology offers the potential to address questions at a scale and with a focus that cannot be undertaken by any single scientific community, agency, or sector.
- Providing a framework for different communities to work together will lead to synergies and new approaches that no single community could have achieved alone.
- A broad array of programs to identify, support, and facilitate biology research exists in the federal government but value is being lost by not integrating these efforts.
- Interagency insight and oversight is critical to support the emergence and growth of the New Biology Initiative. Interagency leadership will be needed to oversee and coordinate the implementation of the initiative, evaluate its progress, establish necessary working subgroups, maintain communication, guard against redundancy, and identify gaps and opportunities for leveraging results across projects.

Recommendation 2:
The committee recommends that the national New Biology Initiative be an interagency effort, that it have a timeline of at least 10 years, and that its funding be in addition to current research budgets.

Finding 3

- Information is the fundamental currency of the New Biology.
- Solutions to the challenges of standardization, exchange, storage, security, analysis, and visualization of biological information will multiply the value of the research currently being supported across the federal government.
- Biological data are extraordinarily heterogeneous and interrelating various bodies of data is currently precluded by the lack of the necessary information infrastructure.
- It is critical that all researchers be able to share and access each others' information in a common or fully interactive format.
- The productivity of biological research will increasingly depend on long-term, predictable support for a high-performance information infrastructure.

Recommendation 3

The committee recommends that, within the national New Biology Initiative, priority be given to the development of the information technologies and sciences that will be critical to the success of the New Biology.

Finding 4

- Investment in education is essential if the New Biology is to reach its full potential in meeting the core challenges of the 21st century.
- The New Biology Initiative provides an opportunity to attract students to science who want to solve real-world problems.
- The New Biologist is not a scientist who knows a little bit about all disciplines, but a scientist with deep knowledge in one discipline and a "working fluency" in several.
- Highly developed quantitative skills will be increasingly important.
- Development and implementation of genuinely interdisciplinary undergraduate courses and curricula will both prepare students for careers as New Biology researchers and educate a new generation of science teachers well versed in New Biology approaches.
- Graduate training programs that include opportunities for interdisciplinary work are essential.
- Programs to support faculty in developing new curricula will have a multiplying effect.

Recommendation 4

The committee recommends that the national New Biology Initiative devote resources to programs that support the creation and implementation of interdisciplinary curricula, graduate training programs, and educator training needed to create and support New Biologists.

References

Altshuler, D., Daly, M. J., & Lander, E. S. (2008). Genetic mapping in human disease. *Science, 322*(5903), 881-888.

Backhed, F., Ley, R. E., Sonnenburg, J. L., Peterson, D. A., & Gordon, J. I. (2005). Host-bacterial mutualism in the human intestine. *Science, 307*(5717), 1915-1920.

Blakeslee, S. (2008, January 15). Monkey's Thoughts Propel Robot, a Step That May Help Humans. *New York Times.*

Buchbinder, S. P., Mehrotra, D. V., Duerr, A., Fitzgerald, D. W., Mogg, R., Li, D., et al. (2008). Efficacy assessment of a cell-mediated immunity HIV-1 vaccine (the Step Study): a double-blind, randomised, placebo-controlled, test-of-concept trial. *Lancet, 372*(9653), 1881-1893.

Bureau of Transportation Statistics, U.S. Department of Transportation (2009). *National Transportation Statistics.* Retrieved 07/23/2009, from http://www.bts.gov/publications/national_ transportation_statistics/#chapter_4.

Burke, D., Bunnell, J., Collins, J., Morse, S., Riley, L., Russek-Cohen, E., & Trtanj, J. (2005). *Review of the Joint National Institutes of Health/National Science Foundation Ecology of Infectious Disease Program.* Retrieved 8/14/2009 from http://www.fic.nih.gov/programs/research_grants/ ecology/eid_review2005.pdf.

Carthew, R. W., & Sontheimer, E. J. (2009). Origins and mechanisms of miRNAs and siRNAs. *Cell, 136*(4), 642-655.

Census Bureau, U.S. Department of Commerce (2008). International Data Base 2008 Update. Retrieved 8/21/2009, from http://www.census.gov/ipc/www/idb/worldpopgraph.php.

Chen, D., Du, W., Liu, Y., Liu, W., Kuznetsov, A., Mendez, F. E., et al. (2008). The chemistrode: A droplet-based microfluidic device for stimulation and recording with high temporal, spatial, and chemical resolution. *Proceedings of the National Academy of Sciences, 105*(44), 16843-16848.

Chivian, E., & Bernstein, A. (2008). *Sustaining life: how human health depends on biodiversity.* Oxford, UK; New York: Oxford University Press.

Dahan, M., Levi, S., Luccardini, C., Rostaing, P., Riveau, B., & Triller, A. (2003). Diffusion Dynamics of Glycine Receptors Revealed by Single-Quantum Dot Tracking. *Science, 302*(5644), 442-445.

DOE Joint Genome Institute (2009). *Who We Are.* Retrieved 7/15/2009, from http://www.jgi.doe. gov/whoweare/whoweare.html.

Egli, D. B. (2008). Comparison of corn and soybean yields in the United States: Historical trends and future prospects. *Agronomy Journal, 100*(3), S79-S88.

EIA (Energy Information Administration), U.S. Department of Energy (2007). *Annual Energy Review*. Retrieved 5/4/2009, from http://www.eia.doe.gov/aer/.

Eid, J., Fehr, A., Gray, J., Luong, K., Lyle, J., Otto, G., et al. (2009). Real-Time DNA Sequencing from Single Polymerase Molecules. *Science, 323*(5910), 133-138.

EPSRC (Engineering and Physical Sciences Research Program) (2009). *What is a Sandpit?* Retrieved 12/10/2008, from http://www.epsrc.ac.uk/ResearchFunding/Opportunities/Networking/IDEASFactory/WhatIsASandpit.htm.

FAO (Food and Agriculture Organization of the United Nations) (2008). *The State of Food Insecurity in the World*. Rome: Author.

Fehr, W. R. (2007). Breeding for Modified Fatty Acid Composition in Soybean. *Crop Science, 47*(Supplement 3), S-72-87.

Fishel, R., & Kolodner, R. D. (1995). Identification of mismatch repair genes and their role in the development of cancer. *Current Opinion in Genetics & Development, 5*(3), 382-395.

Gao, X. H., Cui, Y. Y., Levenson, R. M., Chung, L. W. K., & Nie, S. M. (2004). In vivo cancer targeting and imaging with semiconductor quantum dots. *Nature Biotechnology, 22*(8), 969-976.

H. John Heinz III Center for Science, Economics, and the Environment. (2002). *The state of the nation's ecosystems: measuring the lands, waters, and living resources of the United States*. Cambridge, UK; New York: Cambridge University Press.

H. John Heinz III Center for Science, Economics, and the Environment. (2008). *The state of the nation's ecosystems 2008: measuring the lands, waters, and living resources of the United States*. Washington, DC: Island Press.

Hagen, J. B. (1992). *An entangled bank: The origins of ecosystem ecology*. New Brunswick, NJ: Rutgers University Press.

Handelsman, J., Ebert-May, D., Beichner, R., Bruns, P., Chang, A., DeHaan, R., et al. (2004). Scientific teaching. *Science, 304*(5670), 521-522.

Hejnol, A., Obst, M., Stamatakis, A., Ott, M., Rouse, G. W., Edgecombe, G. D., Martinez, P., Baguña, J., Bailly, X., Jondelius, U., Wiens, M., Müller, W. E. G., Seaver, E., Wheeler, W. C., Martindale, M. Q., Giribet, G. & Dunn, C. W. (2009) Assessing the Root of Bilaterian Animals with Scalable Phylogenomic Methods. *Proceedings of the Royal Society B: Biological Sciences*, 2009 Sep 16 [Epub ahead of print].

Hill, S. T. (2006). *Science and Engineering Doctorate Awards: 2005 - Detailed Statistical Tables: NSF 07-305: December 2006*: National Science Foundation.

Houghton, J. T., & Intergovernmental Panel on Climate Change. Working Group I. (2001). *Climate change 2001 : the scientific basis : contribution of Working Group I to the third assessment report of the Intergovernmental Panel on Climate Change*. Cambridge, UK; New York: Cambridge University Press.

IEA (International Energy Agency) (2008). *World Energy Outlook 2008*. Paris, France: Author.

Institute of Medicine (U.S.). (2008). *From molecules to minds: challenges for the 21st century: workshop summary*. Washington, DC: The National Academies Press.

Leis, A., Rockel, B., Andrees, L., & Baumeister, W. (2009). Visualizing cells at the nanoscale. *Trends in Biochemical Sciences, 34*(2), 60-70.

Maloy, S., & Schaechter, M. (2006). The era of microbiology: a Golden Phoenix. *International Microbiology, 9*(1), 1-7.

Martin, V. J. J., Pitera, D. J., Withers, S. T., Newman, J. D., & Keasling, J. D. (2003). Engineering a mevalonate pathway in Escherichia coli for production of terpenoids. *Nature Biotechnology, 21*(7), 796-802.

Millennium Ecosystem Assessment (Program) (2005). *Ecosystems and human well-being: Synthesis*. Washington, DC: Island Press.

National Academies. (2004). *Facilitating interdisciplinary research*. Washington, DC: The National Academies Press.

National Academy of Engineering (Producer). (2009). *Grand Challenges for Engineering*. Podcast retrieved from http://www.engineeringchallenges.org/.

National Academy of Engineering. (2005). *Educating the engineer of 2020 : adapting engineering education to the new century*. Washington, DC: The National Academies Press.

National Center for Ecological Analysis and Synthesis (2009). Welcome to NCEAS, from http://www.nceas.ucsb.edu/.

National Center for Biotechnology Information (2009, April 27). GenBank Overview. Retrieved 6/17/2009, from http://www.ncbi.nlm.nih.gov/Genbank/.

National Center for Research Resources (2009). Advisory Council Meeting Minutes, February 12, 2009. Retrieved 7/1/2009, from http://www.ncrr.nih.gov/about_us/advisory_council/minutes/20090212.asp.

National Research Council (1989). *Opportunities in biology*. Washington, DC: National Academy Press.

National Research Council (1990). *Fulfilling the promise: Biology education in the nation's schools*. Washington, DC: National Academy Press.

National Research Council (1999). *Transforming undergraduate education in science, mathematics, engineering, and technology*. Washington, DC: National Academy Press.

National Research Council (2000a). *Ecological indicators for the nation*. Washington, DC: National Academy Press.

National Research Council (2000b). *How people learn: Brain, mind, experience, and school* (Expanded ed.). Washington, DC: National Academy Press.

National Research Council (2002). *Learning and understanding: Improving advanced study of mathematics and science in U.S. high schools*. Washington, DC: The National Academies Press.

National Research Council (2003a). *Bio 2010: Transforming undergraduate education for future research biologists*. Washington, DC: The National Academies Press.

National Research Council (2003b). *Evaluating and improving undergraduate teaching in science, technology, engineering, and mathematics*. Washington, DC: The National Academies Press.

National Research Council. (2003c). *Large-scale biomedical science: exploring strategies for future research*. Washington, DC: The National Academies Press.

National Research Council (2005). *How students learn: Mathematics in the classroom*. Washington, DC: The National Academies Press.

National Research Council (2006). *America's lab report: Investigations in high school science*. Washington, DC: The National Academies Press.

National Research Council (2007a). *Understanding Interventions that Encourage Minorities to Pursue Research Careers: Summary of a Workshop*. Washington, DC: The National Academies Press.

National Research Council (2007b). *Earth Science and Applications from Space.* Washington, DC: The National Academies Press.

National Research Council (2008). *Achievements of the National Plant Genome Initiative and new horizons in plant biology*. Washington, DC: The National Academies Press.

National Research Council (2009). *The Role of the Life Sciences in Transforming America's Future: Summary of a Workshop.* Washington, DC: The National Academies Press.

NCBI (2009, 4/27/2009). *GenBank Overview* Retrieved 6/17/2009, from http://www.ncbi.nlm.nih.gov/Genbank/.

NCEAS (National Center for Ecological Analysis and Synthesis) (2009). *Welcome to NCEAS.* Retrieved 7/10/2009, from http://www.nceas.ucsb.edu/.

NIH (National Institutes of Health) (2009, June 1). *NIH Roadmap for Medical Research.* Retrieved 6/17/2009, from http://nihroadmap.nih.gov/.

NSTC (National Science and Technology Council) (2003). *National Plant Genome Initiative: 2003–2008.* Retrieved 8/21/09 from http://www.ostp.gov/pdf/npgi2003_2008.pdf.

O'Malley, R., Marsh, A. S., & Negra, C. (2009). Closing the Environmental Data Gap. *Issues in Science and Technology, 25*(3), 69-74.

Pacific Northwest National Laboratory (2009). *Bio-based Product Research*. Retrieved 6/18/2009, from http://www.pnl.gov/biobased/.

Pfund, C., Miller, S., Brenner, K., Bruns, P., Chang, A., Ebert-May, D., et al. (2009). Summer Institute to Improve University Science Teaching. *Science, 324*(5926), 470-471.

Sanchez, J.C., Mahmoudi, B., DiGiovanna, J., & Principe, J. C. (2009). Coadaptive brain-machine interface via reinforcement learning. *IEEE Transactions on Biomedical Engineering, 56*(1), 54-64.

Schwenk, K., Padilla, D. K., Bakken, G. S., & Full, R. J. (2009). Grand challenges in organismal biology. *Integrative and Comparative Biology, 49*(1), 7-14.

Shendure, J., & Ji, H. L. (2008). Next-generation DNA sequencing. *Nature Biotechnology, 26*(10), 1135-1145.

Shiver, J. W., Fu, T.-M., Chen, L., Casimiro, D. R., Davies, M.-E., Evans, R. K., et al. (2002). Replication-incompetent adenoviral vaccine vector elicits effective anti-immunodeficiency-virus immunity. *Nature, 415*(6869), 331-335.

Steen, L. A. (2005). *Math and Bio 2010: Linking undergraduate disciplines.* Washington, DC: Mathematical Association of America.

The White House, Office of the Press Secretary (April 27, 2009). *Remarks by the President at the National Academy of Sciences Annual Meeting. The White House Press Release.* Retrieved 8/1/2009, from http://www.whitehouse.gov/the_press_office/Remarks-by-the-President-at-the-National-Academy-of-Sciences-Annual-Meeting/.

Turnbaugh, P. J., Hamady, M., Yatsunenko, T., Cantarel, B. L., Duncan, A., Ley, R. E., et al. (2009). A core gut microbiome in obese and lean twins. *Nature, 457*(7228), 480-U487.

USDA (U.S. Department of Agriculture) (2009). *U.S. domestic corn use.* Retrieved 7/10/2009, from http://www.ers.usda.gov/Briefing/Corn/Gallery/Background/CornUseTable.html.

USDA & University of California, B. (2009). *The California Institute for Quantitative Biosciences*, from http://research.chance.berkeley.edu/page.cfm?id=56.

Woese, C. R., & Fox, G. E. (1977). Phylogenetic structure of prokaryotic domain - primary kingdoms. *Proceedings of the National Academy of Sciences of the United States of America, 74*(11), 5088-5090.

Woese, C. R., & Goldenfeld, N. (2009). How the Microbial World Saved Evolution from the Scylla of Molecular Biology and the Charybdis of the Modern Synthesis. *Microbiology and Molecular Biology Reviews, 73*(1), 14-21.

Appendix A

Statement of Task

An ad hoc committee will examine the current state of biological research in the United States and recommend how best to capitalize on recent technological and scientific advances that have allowed biologists to integrate biological research findings, collect and interpret vastly increased amounts of data, and predict the behavior of complex biological systems. Among the questions the committee may address are:

• What fundamental biological questions are ready for major advances in understanding? What would be the practical result of answering those questions? How could answers to those questions lead to high impact applications in the near future?

• How can a fundamental understanding of living systems reduce uncertainty about the future of life on earth, improve human health and welfare, and lead to the wise stewardship of our planet? Can the consequences of environmental, stochastic or genetic changes be understood in terms of the related properties of robustness and fragility inherent in all biological systems?

• How can federal agencies more effectively leverage their investments in biological research and education to address complex problems across scales of analysis from basic to applied? In what areas would near term investment be most likely to lead to substantial long-term benefit and a strong, competitive advantage for the United States? Are there high-risk, high pay-off areas that deserve serious consideration for seed funding?

• Are new funding mechanisms needed to encourage and support cross-cutting, interdisciplinary or applied biology research?

• What are the major impediments to achieving a newly integrated biology?

• What are the implications of a newly integrated biology for infrastructural needs? How should infrastructural priorities be identified and planned for?

• What are the implications for the life sciences research culture of a newly integrated approach to biology? How can physicists, chemists, mathematicians and engineers be encouraged to help build a wider biological enterprise with the scope and expertise to address a broad range of scientific and societal problems?

• Are changes needed in biology education—to ensure that biology majors are equipped to work across traditional subdisciplinary boundaries, to provide biology curricula that equip physical scientists and engineers to take advantage of advances in biological science, and to provide nonscientists with a level of biological understanding that gives them an informed voice regarding relevant policy proposals? Are alternative degree programs needed or can biology departments be organized to attract and train students able to work comfortably across disciplinary boundaries?

The subgroup of the committee will organize a Biology Summit to garner input from a broad spectrum of stakeholders—government and private agencies that fund biological research, the biotech and pharmaceutical industries, universities and medical schools—to consider barriers to progress and to highlight exciting new areas of research that cross traditional disciplinary boundaries. The Summit's proceedings will be published as a separate, type-3 workshop report. In a subsequent consensus report, the committee will recommend actions that federal policy makers can take to ensure that the United States takes the lead in the emergence of a biological science that will support a higher level of confidence in our understanding of living systems, thus reducing uncertainty about the future, contributing to innovative solutions for practical problems, and allowing the development of robust and sustainable new technologies. The committee will not make specific budgetary or government organizational recommendations.

Appendix B

Workshop Agenda

December 3, 2008
1200 New York Avenue NW
Washington, D.C. 20005

8:30 a.m. Welcome from the President
            Ralph Cicerone, President of the National Academy of
            Sciences
8:45 a.m. James P. Collins
            Assistant Director for Biological Sciences, National Science
            Foundation
9:15 a.m. Raymond L. Orbach
            Undersecretary for Science in the United States Department
            of Energy
9:45 a.m. Thomas R. Cech
            President of the Howard Hughes Medical Institute
10:15 a.m. David T. Kingsbury
            Chief Program Officer for Science, Gordon and Betty Moore
            Foundation

10:45 a.m. Break

11:00 a.m. Susan Hockfield
            President of the Massachusetts Institute of Technology
11:30 a.m. Susan Desmond-Hellmann
            President of Product Development, Genentech
12:00 p.m. Harold Varmus
            President of the Memorial Sloan-Kettering Cancer Center

12:30 p.m.   Panel Discussion

1:00 p.m.   Lunch Break

2:15 p.m.   Cynthia Kenyon
             American Cancer Society Professor, UCSF
2:45 p.m.   Lucy Shapiro
             Ludwig Professor of Cancer Research, Stanford University

3:15 p.m.   Break

3:30 p.m.   Robert Fraley
             Executive Vice President and Chief Technology Officer,
             Monsanto
4:00 p.m.   Elias A. Zerhouni
             Former Director of the National Institutes of Health

4:30 p.m.   Panel Discussion

5:00 p.m.   Closing Remarks
             Keith Yamamoto
             Chairman, Board on Life Sciences, National Research Council